A VOICE FOR THE DEAD

A VOICE FOR THE DEAD

A Forensic Investigator's Pursuit of the Truth in the Grave

JAMES E. STARRS

WITH KATHERINE RAMSLAND

G. P. Putnam's Sons *New York*

G. P. PUTNAM'S SONS

Publishers Since 1838

Published by the Penguin Group
Penguin Group (USA) Inc., 375 Hudson Street, New York, New York 10014, USA • Penguin Group
(Canada), 10 Alcorn Avenue, Toronto, Ontario, Canada M4V 3B2 (a division of Pearson Penguin
Canada Inc.) • Penguin Books Ltd, 80 Strand, London WC2R 0RL, England • Penguin Ireland,
25 St Stephen's Green, Dublin 2, Ireland (a division of Penguin Books Ltd) • Penguin Group
(Australia), 250 Camberwell Road, Camberwell, Victoria 3124, Australia (a division of Pearson Australia
Group Pty Ltd) • Penguin Books India Pvt Ltd, 11 Community Centre, Panchsheel Park,
New Delhi–110 017, India • Penguin Group (NZ), Cnr Airborne and Rosedale Roads, Albany,
Auckland 1310, New Zealand (a division of Pearson New Zealand Ltd) • Penguin Books (South
Africa) (Pty) Ltd, 24 Sturdee Avenue, Rosebank, Johannesburg 2196, South Africa

Penguin Books Ltd, Registered Offices: 80 Strand, London WC2R 0RL, England

The authors acknowledge permission to reprint from the following:
"At Last the Secret Is Out" by W. H. Auden, from *The Ascent of F6* by W. H. Auden
and Christopher Isherwood. Copyright © 1936 by Wystan Hugh Auden and
Christopher Isherwood, Renewed. Reprinted by permission of Curtis Brown, Ltd.
"Truth" by John Masefield. Copyright 1935 by John Masefield. Reprinted by permission of
The Society of Authors as the Literary Representative of the Estate of John Masefield.

Library of Congress Cataloging-in-Publication Data

Starrs, James E.
A voice for the dead: a forensic investigator's pursuit of the truth in
the grave/by James E. Starrs with Katherine Ramsland.
p.   cm.
Includes bibliographical references and index.
ISBN 0-399-15225-3
1. Criminal investigation—Case studies.   2. Homicide investigation—Case studies.
3. Exhumation—Case studies.   4. Forensic sciences.
I. Ramsland, Katherine M., date.   II. Title.
HV8073.5.S75   2005   2004053463
363.25'9523—dc22

Printed in the United States of America
1   3   5   7   9   10   8   6   4   2

This book is printed on acid-free paper. ♾

BOOK DESIGN BY MEIGHAN CAVANAUGH

TO THE ONES WHOSE VOICES

HAVE NOT YET BEEN HEARD

BUT DESERVE TO BE

ACKNOWLEDGMENTS

W. H. Auden was right as rain when he proposed that "words have no words for words that are not true."

Some words, like "sorry," come all too trippingly to the tongue, putting the imprint of insincerity to the statement. Other words, like the commonplace and perfunctory greeting "how are you?" are not intended to instigate a conversation or even a truthful response. Indeed, even the "grand day, praise God" greeting regularly encountered from passersby in Irish country laneways sometimes seems too spontaneous and formulaic to be genuine, especially when dark clouds are threatening an imminent downpour.

Possibly the worst offender among words that are altogether likely to be considered to be, at first blush, untrue is "thank you." To counter that possibility, modifiers are often called to the task of giving a "thank-you" a more positive and convincing stature, such as the declaration "thank you very much" or by going overboard in the effort to eschew ingratitude with a hearty, almost backslapping "thank you from the bottom of my heart."

The short of it is that an acknowledgment, to be a sincere and genuine expression of appreciation, is deucedly difficult to articulate in so many words. Yet the effort is worth the risk, for to forgo the much-deserved recognition of the efforts of the numerous contributors to the success of my various exhumation projects would be to pass them by, as Shakespeare's Achilles rightly complained, "as misers do by beggars," neither giving "good word nor look," signifying meanly that their "deeds (are) forgot" or, worse yet, not worth memorializing. But how to express my gratitude with conviction is the nub of the problem.

The fact that my exhumation projects now number in the double digits and that each project has been multidisciplinary in nature, staffed by scientists and nonscientists totaling nearly a hundred different persons, leaves me with no alternative other than to express my abiding and forthright appreciation en masse to all those persons who have volunteered to involve themselves in one or more of my exhumation activities. This broadly based and all-encompassing recognition is not intended to downplay the work performed at my behest by any of those in the many diverse forensic scientific and nonscientific disciplines that empowered my projects. The forensic scientists came from such differing scientific pursuits as pathology, odontology, radiography, radiology, engineering, geophysics, anthropology, archaeology, firearms and toolmark identifications, pharmacology, toxicology, document examiners, molecular biology, chemistry, microscopy, and entomology, as well as all those not mentioned which should have been.

My warmest appreciation is extended also to the lawyers whose legal skills jump-started many of the projects and kept them from legal waywardness throughout their duration. Others outside the disciplines of science who were more than instrumental in shouldering the burdens of these projects were the historians, the photographers, both still and video, and the funeral directors who graciously gave us the sites we so desperately needed for our post-exhumation analyses.

And, lest they be forgotten, the relatives, descendants, and friends of the twenty-five persons exhumed by the multidisciplinary teams must be singled out for a very special vote of gratitude for the trust they reposed in us and for their steely refusal to allow the record of the cause and manner of the death or the location of the burial site of their loved ones to go unexplored, no matter the length of time that had elapsed since their death.

My teams could not have functioned as effectively as they did without the able and sturdy support of innumerable staff members, including secretaries, research assistants, documentarians, and others who provided a logistical uplift in the form of the necessary recordation, the provision of supplies and food to document and to shore up the infrastructure of each and every venture. Each and every one of them deserves and has my fulsome gratitude.

Just as every detail of an exhumation requires planning and decision-making, so too is an acknowledgment such as this beset by imperatives involving hard choices. I could vainly attempt to name the names of all those who aided in my endeavors but that would be monstrously unrewarding, for, sure as there are senior moments, I would leave a person unnamed both to my shame and to their embarrassment. There just is no way to ensure against the omission in such a list of names of one or more, most likely the latter, persons who are preeminently deserving of being singled out in an acknowledgment crediting their participation.

Furthermore, even though this book concentrates on a number of high-profile exhumations which were consummated, as well as a selection of others which have been either temporarily or permanently put on hold, still there are other exhumations, such as those of Samuel Swan, Carl Williams, Thomas Traylor, and Samuel Washington, which were the stuff of a raconteur's roundtable but were beyond the reach of this modest-sized authorial and publishing venture. However, this is a fitting occasion to give praise to the persons who contributed to all of my exhumations, whether encapsulated in this book or not.

As this book's index will readily disclose, the names of certain individuals are not shrouded in the anonymity that surrounds others whose participation is acclaimed in this acknowledgment. Persons were named in the text when the vignettes or incidents in which they were closely involved could not be compellingly told without naming them.

My grandest good fortune was in being surrounded by persons who would not be taken in by old saws and canards concerning the mysteries we jointly explored. They came to each project with an observant open-mindedness to be expected only of the most disciplined and conscientious investigators. They would not prejudge the merits of the various exhumations, wrapped as most were in myth and legend. Unlike the newspaper editor in the John Ford movie *The Man Who Shot Liberty Valance* who, with H. L. Mencken–like cynicism, proclaimed, "When the legend becomes fact, print the legend," the members of my teams were disposed to dispel the legends and to replace them with the truth as known only by our scientific data and investigative acuity. It was a singular blessing for me to work with persons of such untrammeled scientific ideals and undeterred moral stamina. My choices of teammates left me in constant awe of their vigor, perspicacity, and resounding good humor.

The one person, so far unnamed, who cannot be left unnamed is my wife of fifty years, Barbara Alice Starrs, who often labored by my side uncomplainingly, even though she wondered aloud on many occasions about the whys and wherefores of my various scientific undertakings, including the fashioning of this book. She was my gentle Jacques, always there to remind me that "all the world's a stage and all the men and women merely players," having "their exits and their entrances." Such an outlook helped keep my head on tight, my goals in perspective, and, it is devoutly hoped, my words true.

CONTENTS

A VOICE FOR THE DEAD

INTRODUCTION

Desiderata for an Exhumation

The times have been
That, when the brains were out, the man would die,
And there an end; but now they rise again . . .

SHAKESPEARE, *MACBETH*, ACT III, SCENE 4, LL. 78–81

Exhumations, don't you know, resemble parades. Both happen infrequently and both occur only as a very special event for a very special purpose. They differ in that the prominence of exhumations nowadays has far outpaced the appeal of parades.

Josef Kanat, alias Wolfgang Gerhard, has been identified as Josef Mengele, Nazi Germany's Angel of Death, but only after an exhumation in Brazil. Lee Harvey Oswald was in fact buried in Rosehill Cemetery in Dallas, Texas, and not a stand-in Russian agent. We know that only because of an exhumation conducted in Texas in 1981. Zachary Taylor, former president of the United States, did not die as a result of homicidal poisoning, or so his exhumation from his crypt in a Kentucky cemetery has concluded. But there are still confounding questions about the deaths of other prominent, or even less notable, persons—at least to the world at large. Should the answers or, at least, clarification be sought in exhumations of their remains?

Is William Casey, former director of the CIA, really dead of a brain tumor, or was his purported death at the height of the investigation into the Iran-Contra affair merely a cover to enable him to be spirited off to a place of sanctuary until the heat of the investigation had dwindled? Should an exhumation be authorized to determine whether Casey's remains are truly buried in his assumed final resting place?

Should Edgar Allan Poe be exhumed from his burial place in the Westminster Church yard in Baltimore to determine whether he died of rabies or alcoholism or a fatal assault or, less likely, a hospital-induced mischance? Should the likelihood of success be the principal motivating factor? How about the willingness of Poe's descendants to consent to an exhumation? Is the likely condition of the remains a signal factor to reckon with? What legal requirements must be satisfied to gain approval for such an exhumation?

Missourians and Kentuckians alike have been at loggerheads for centuries over whether Daniel Boone and his wife are buried in Missouri or Kentucky. Some would argue that he could be, as only Daniel Boone might be, in both places at the same time—at least part of him in one state and parts in the other. Would an exhumation be in order to stop this warring between the states?

Like parades, there are good exhumations and there are bad exhumations, as viewed from the historical controversy surrounding them and the scientific values in conducting them. But regardless of their merit, exhumations, like parades, are spectacles, occasionally even extravaganzas, which captivate the public and become conversation pieces. But exhumations cannot be all high-flying batons and gaudy floats if they are to prove their scientific mettle.

At the outset certain distinctions should be borne in mind, in part to distinguish that which is an exhumation from that which is not. Throughout this book, I speak of exhumations as if I am either oblivious to or neglectful of the differences between an exhumation, also known as a disinterment, and an excavation, also known as an archaeological dig. There is no fence-post line standing out boldly and plainly as a fixed demarcation between an exhumation and an excavation.

Whereas excavations tend to be archaeological in character, searching more for artifacts and structures reflecting past human colonization than for human remains, still my efforts to locate the remains of buried Confederate soldiers at the Gettysburg, Pennsylvania, battlefield field hospital where they were buried was certainly more excavation than exhumation, even though my objective was to find those documented remains on this forty-plus-acre, much-farmed-over, and rock-strewn terrain. On the other hand, the continuing digging at the Jamestown site in colonial Virginia was in its beginning and is still archaeological in nature, even though human skeletal remains have been recovered there.

It is not temporal factors that separate an excavation from an exhumation. Even though exhumations generally are of burials of more recent vintage, as was my exhumation of Chief Petty Officer Thomas Traylor in 2003 in Big Piney, Wyoming, he having died in 2002, still Jesse James had been dead for 107 years when I exhumed his remains in 1989.

And Samuel Washington had been said to be in his grave at his Harewood estate outside Charles Town (named after another brother of George Washington), West Virginia, in excess of 200 years (he died in 1781) when I sought to locate his remains in 1999; that enterprise was both an excavation and an exhumation. Samuel, supposedly resting in an unmarked grave in a walled family cemetery at Harewood, was the target of our efforts, but the means we employed to that end were particularly of the archaeological kind. Whether an exhumation or an excavation, strictly so-called, we could say as the end product of our digging that Samuel was not buried within the confines of the walled cemetery where the received tradition said he was.

Excavators tend to use the archaeological tools of variations in soil coloration (something not taken into account on a Munsell chart, which describes the color of soil by hue, value, and chroma), resulting from admixture of soil zonation levels, as well as soil subsidence from the settling of soil at a grave site or from the collapse of a coffin "six feet under" and gentle probing with metal rods, to detect the softer soil resulting from a burial.

But those rudimentary markers for the presence of a burial site have also been put to the task by me in exhumations such as that of Jesse James, where the mixed soil coloration gave me a baseline for the size and location of Jesse's grave at Mount Olivet Cemetery in Kearney, Missouri. I was, thereby, saved from disturbing a family member's adjoining grave, which would have violated the spirit of the court order granting me the authority to exhume Jesse.

The boundaries between exhumations and excavations being neither neat nor tidy, I have in this book used the terms interchangeably, preferring, however, to describe my work as an exhumation since the exclusive objective for me is finding human remains.

I have avoided derogatory terms like "grave digger" and "grave robber," since I claim no bloodlines or other connection, implied or express, linking me to Edinburgh, Scotland's, odious Burke and Hare, who were grave robbers turned murderers. Burke and Hare were engaged in a commercial venture designed for them to profit from their providing cadavers for the medical experimentation of Edinburgh's Dr. Knox. Unlike Burke and Hare, I am not of the "have shovel, will dig" stripe, nor am I moved by the desire for lucre. Instead I insist that an exhumation be a rare occurrence, and then only within the limits of strictly applied criteria.

There are exhumations and then there are exhumations—of the far less responsible and creditable kind. One man called to ask me to exhume his grandfather. When I asked, "For what purpose?" he replied, "So that I can get to know him." Stunned by this frank but confusing answer, I asked how an exhumation would enable him to know his grandfather. His reply almost knocked me back out of my desk chair. "An exhumation will give me the chance to touch his bones," he stated unabashedly. My riposte was quick and to the point. "Mister," I said, "you don't need a forensic scientist, you need a psychiatrist." End of telephone conversation.

Let it be known that I turn down many more exhumation requests than I ultimately agree to sponsor. A stranger-than-thou request came to me from faraway Manila some years ago asking me to assist in locating the buried treasure of Japanese general Yamashita, the man General McArthur ousted from the Philippines during the Second World War. When I was informed that I would be provided two bodyguards for my protection during my stay, I decided I had less risky ways to spend my time, especially since there was only a pot of supposed gold at the end of this journey, not human remains—unless they were mine. As my wife regularly quips, "Jim will stop doing these exhumations when people *start* paying him to do them. That's his contrariness at work."

The apparent voguishness of exhumations of late has failed to distinguish between exhumations that are justifiable and those that are not. Or, to put it in unvarnished terms, some exhumations are unwarranted while others are entirely proper and scientifically supportable. The rub lies in making the sometimes thinly lined distinction between exhumations for just cause and those that have no such legitimacy.

One thing is pellucidly clear on the subject of whether to exhume or not to exhume. Exhumations are based on decisions that are never clear-cut or simple, either in the making or in the execution. Mock editorializing about the need for an "Exhumed Presidents Society" to oversee the exhumation of presidents and others of such stature in history serves nothing but a Swiftian, satirical purpose when what is needed is a sober and conscientious statement and analysis of standards, or at least guidelines, for an exhumation.

I've seen some exhumations that are irresponsible attempts to disturb the dead for the sake of proving a harebrained theory, and I've seen others that are scientifically worthy. Some notable people die surrounded by legends and half-truths, making it legitimate to exhume their remains in an age where science can supply answers to the cause and manner of death, especially if the person in question has historical significance.

Without being conspiracy theorists, we can say that the questions raised on the death of an individual can be many and varied. For example, why did three medical doctors decide not to autopsy the remains of J. Edgar Hoover, a man with many enemies and no history of medical ailments? Shouldn't we find out more through an exhumation? In another historical matter that has brought legions of historians to divergent opinions on the subject, shouldn't the remains of Meriwether Lewis be unearthed and examined in order to determine whether his untimely death in 1809 in Tennessee at the age of thirty-five was a suicide or a homicide on the then lonely and forbidding Natchez Trace Indian Trail? Is the recommendation for an exhumation of a coroner's inquest, 160 descendants of Lewis, and the governors of Missouri, Tennessee, and Virginia not enough to satisfy the concerns of the National Park Service, which lays claim to ownership of his middle Tennessee grave site?

There is a difference between accidental discoveries, such as coming across skeletal remains in a sunken ship, such as the Confederate submarine, the Hurley, or at a Colonial colony like Jamestown, and intentional discoveries, in which we set out to learn with the aid of science what happened to the deceased. For example, the exhumations in the Arctic Circle involving people who died there of smallpox were intentional, for the purposes of obtaining a sample of the virus so it could be cultured, and an antidote developed for storage in case it should be needed one day to prevent an epidemic.

In the kind of exhumation that captivates me, the intentional kind, a scientific team must be assembled and proper procedures followed with a worthwhile goal in mind for what is to be expected to be discovered.

Even so, the decision to exhume should not be a slapdash judgment without the seasoning of intense aboveground investigations and deliberation. Furthermore, the necessary permissions from the courts, if required, or from controlling agencies, such as the West Virginia Division of Culture & History, which approved and oversaw my exhumations/excavations for the remains of Samuel Washington at Harewood in Jefferson County, West Virginia, are a curb on unfounded digging of graves.

And, naturally, attention must be given to obtaining funding for the project. No one wakes up in the morning with an exhumation in mind and then puts everything together in a day. While it can be an exciting and rewarding endeavor, there are obstacles to overcome and questions to answer in the process in order to ensure that the nonmonetary rewards ultimately outweigh the deficits. Thus, for an exhumation to be warranted, I require minimally that the following conditions be fulfilled:

- There must be a significant dispute to which science can make a contribution.
- Some type of new scientific method must be available that was not present at the time the issue to be resolved arose.
- There must be more than a likelihood that the remains will be in sufficient condition to be analyzed, which involves a variety of tasks such as testing the soil pH and mineral content where the remains are buried, determining the method and depth of burial, knowing the time since burial, gathering climatic and environmental data, and forestalling the disturbance of adjacent graves.

One of the first and most taxing tasks is to collect all available historical documentation, from court records to books and newspapers to witness reports. Reliance may even be placed on museum displays, videotapes, or personal records. If burial records and prior autopsy documentation are available, they should be reviewed as well as any documentation from law enforcement (if relevant) at the time of death of the subject under investigation. One should be aware that state statutes concerning access to records of autopsies range from those where not even relatives can freely obtain a copy to those, like Texas's, where such documents are a matter of public record, available for any Johnny-come-lately to peruse.

Then legal counsel should be consulted, always remembering that lawyers expect to be paid and to gain public exposure from your project. Further, consent from relatives should be obtained in formal, notarized documents, one for the exhumation and another for the scientific follow-up. As to the consent form for the scientific work product, it is important to be very broad and open-ended, for there may be destructive testing and other testing aspects that were not contemplated nor could have been at the outset.

Of course, a court order or administrative agency permit may be legally necessary. If not, avoid these legalities like the plague. Those who thrive on going to court in such matters are only those who enjoy licking honey from a razor blade. The courts tend to look at exhumations as downright outrageous in the first instance and must be convinced otherwise. It is as if you are presenting a particularly loathsome scene from one or more of the Texas Chain Saw Massacre flicks into the sober, oak-embedded austerity of the courtroom. In court it is an uphill battle, even with the family's backing.

Prior to any exhumation, the eventual disposition of the remains must be determined, whether they will be reinterred or cremated. The total anticipated costs should also be calculated, and decisions made about whether to photograph the exhumation and autopsy process or to involve the press. I myself would avoid any exhumation where full documentation by videography and still photography is not the gold standard.

Since the remains are usually transported to a lab for analysis, the method and means of transport must be decided. Skeletonized material (known in the scientific vernacular as a skinny) has one set of requirements, while soft-tissue remains (known as a stinky) have quite another. Each state, from Maine to Hawaii, also has its own regulations regarding the handling of the dead.

It may take several preliminary visits to the site before final decisions are made about costs, team personnel, and the tools needed. The location of a suitable site for the analysis of the remains is vital to the success of an exhumation.

In this regard, local funeral directors and their statewide associations can open many doors. Indeed, they should be the point persons for any exhumation. All facets of burials, exhumations, and reinterments are their professional *raison d'être,* and they are usually very amenable to having their knowledge plumbed, frequently without cost to the inquirer.

I remember well how graciously and gratuitously Kathy Ryan, Esq., of the Pennsylvania Funeral Directors Association, came to my urgent need in the Scranton, Pennsylvania, exhumation of Samuel Swan. Without her unrelenting assistance, the exhumation would have been doomed and we would never have learned that the neck organs vital to our reassessment of the official suicide-by-ligature-strangulation opinion had not been reburied with the remains after the autopsy and had not been accounted for on the report of the autopsy. In other words, for lack of the neck organs we could not reevaluate whether Samuel Swan was killed by his prison warders or whether he had used his shoelaces in a suicide.

I have not used one penny of taxpayers' money in any of my projects. Ninety-nine percent of the costs are borne by me, personally, or by my nonprofit corporation, Scientific Sleuthing, Inc., which has for twenty-seven years published a subscriber journal called *Scientific Sleuthing Review.* I eschew personal financing from the families of those I exhume so that it will do nothing to impede my objectivity in reaching my conclusions.

Exhuming a body involves very pragmatic concerns relating to the use of a backhoe, cemetery expenses, food and lodging for the team, securing insurance, paying legal fees, developing a security procedure at the exhumation site, and determining the need for extra equipment, such as ground-penetrating radar devices.

The weather is always a consideration, as is the time of day (early-morning exhumations are generally the best, for spectators are at a minimum then) and the availability of the team members. Even the type of experts needed for the team must be carefully evaluated, and some effort made to find people with the right skills who are willing to work for expenses only in a mutually collegial environment.

The coordination of all these activities and aspects of exhumation can be quite burdensome and time-consuming, but are essential to success. Above all, being prepared for the unexpected is mandatory from one end of the project to the other. Louis Pasteur's sage advice in conducting experiments must be the nom de guerre of the project. He is reported to have said: "Fortune favors those with a prepared mind."

What happens at an exhumation is not just about bringing the remains of a person to the surface, it is also about the camaraderie of the team, the odd little adventures that happen along the way, and the wonderful surprises that help us to set history straight. Although scientific reports about our findings are de rigueur, the untold stories omitted from such formal reports are just as much a part of the process, and many will be related in the ensuing chapters of this book.

Exhumations, like life itself, are full of unexpected turnings that can be deucedly perplexing. However, exhumations should never be conducted with surprise in the forefront as their goal and justification. Without the certainty of a sound and substantial basis, regardless of the outcome, an exhumation is never in order.

I was drawn to this work in part by my deep concern that historians were making categorical statements about the deaths of historic figures that were more buncombe, as Mark Twain liked to phrase it, than gospel. A French litterateur once said of historians that "God in all His omnipotence cannot change the past. That's why He created historians." All too often, I have found that historians have overstepped the bounds of their discipline and made pronouncements on scientific matters beyond their ken. Stephen Ambrose followed this erroneous path in his demeaning characterizations of Meriwether Lewis as an alcoholic and drug addict.

Historians speculate about the way certain personages died, while scientists use the more exacting tools from the sciences to get to the root cause of a death. There is clearly at work here a contretemps, or at least a disconnect, between historians and scientists when we should be working in harmony toward the same end—the truthful and accurate reconstruction of past events.

As a pedagogue of long standing, I am also interested in educating the public, law enforcement, and others in the value and importance of the forensic sciences rather than have them rely on speculation, conjecture, and the rack and screw. The exhumations I have orchestrated put the values of forensic science on parade before the public. They are, it is to be hoped, an accurate description of the high place of esteem that forensic science should hold in law and in the afterlife. The tales they reveal are not the sort told on CSI television programs, all glitzy gimmickry and scientific buffoonery. I tell it as it is, warts and all, including mine.

But even then, why do it—particularly me, with the well-established credentials of a law school and forensic science teaching career behind me? That why is the question that has most often confronted me in the exhumations I have spearheaded. Why did I, in my middle years, decide that the forensic sciences of and for the dead were to be my later-life calling, putting my long-standing legal career into a second-best category? By what right did I, a neophyte in the forensic sciences, deign to assemble, coordinate, and direct teams of seasoned scientists in the multidisciplinary tasks involved in an exhumation? I have asked myself the same or similar whys many times over in different contexts in my lifetime.

Why did I voluntarily join the Army in 1950, leaving a successful college scholarship in midstream? Why did I, while an Army enlisted man, unhesitatingly volunteer for duty in wartime Korea? Why did I enroll jointly in college and law school after returning from the Korean War?

Why did I volunteer for civil-rights duty as a lawyer in Mississippi and Alabama in the summer of 1965, leaving my wife and newborn baby daughter, Barbara, to undergo serious neonatal surgery in my absence? Why did I boldly, even brazenly, lead groups of my teenage children and their classmates on cross-country bicycle treks twice during my summer off-teaching months? Are all these apparently unrelated events connected in some way? I think so.

The same two common themes course through them all and link them together. They all involved occasions where risk-taking was prominent, and they all concerned my reaching out to others in need, whether in the land of the living or the dead. I have keenly learned that the only risks worth taking are those that make life more than a dream and offer the promise of fulfilling that dream.

Only in experiencing the risks that I have dared to encounter for the greater good does one truly savor life at its fullest. Thus, the risks—yes, even the scorn—evoked in my orchestrating exhumations are as nothing compared with the satisfaction of knowing that I, and my team members, have, as a well-fashioned unit, done something unique and tangible for the greater good of those whose voices would otherwise have passed unheard, unrecognized, and unrequited. For me that suffices to justify each and every one of my exhumations. Of course, it also helped that my paternal grandfather was a funeral director who imbued me with an overweening curiosity about his work.

The cases covered in the following chapters all meet my requirements for an exploration through an exhumation of persons in historical tales that desperately needed the strength and the illumination of the forensic sciences.

The dead cannot speak for themselves, but science, with my team and me as surrogates for the dead, can give them a voice. The scientist in me demands the truth for the dead. The lawyer in me says they deserve their day in court. In reading these chapters, I hope you will agree that death does not silence the voice of the dead, nor should it.

1. ALFRED G. PACKER

The Colorado Cannibal with a . . . Conscience

The evil that men do lives after them,
The good is oft interred in their bones.

SHAKESPEARE, *JULIUS CAESAR,*
ACT III, SCENE 2, LL. 80–81

Cannibalism is not a subject for the thin-skinned, for it reeks of the grisly, the gory, and the gruesome. Yet the public has always been fascinated with it. Recall its stark presence in classic children's stories such as "Hansel and Gretel" and "Little Red Riding Hood."

From Greek mythology to Hollywood's fictionalizing, the topic of cannibalism has found its place in the proscenium arch, as something for the learned and something for everyman. King Cronus is known to us in the image of Father Time, but he was known to the Greeks as having lived to see his dethronement accomplished in spite of his best efforts. It had been prophesied that Cronus would lose his crown to one of his own children. To ensure himself against such an outcome, he ordered his wife and sister Rhea to deliver each child she bore to him so that he might devour them. Five children were thus cannibalized until Rhea, at last enraged, bore Zeus, her third son, and spirited him away from Cronus to safety. Zeus did come into his own and did oust his father from his realm and took such total control as to become Lord of the Universe.

In our time, the movie theaters get filled to overflowing when cannibalism is on their menu. *Ravenous* is the latest in the lists, with its oddball vampirish take on California's Donner Party episode. Yes, there was cannibalism and murder to go with it in the Donner Party's extreme efforts to survive the relentless snow of a California mountain winter, but no vampires have been reported by the few survivors. Movie producers have even sought to portray cannibalism as a matter for S. J. Perelman's style of humor, sometimes, as in *Gentlemen Don't Eat Poets,* with a modicum of success.

The type of cannibalism that follows extreme measures for survival, such as that of the airplane passengers trapped after their plane crashed in the high Andes, can be the tolerable centerpiece of polite parlor room conversation, and even Shakespeare's Othello recounted how he had enthralled Desdemona with such tales. So it should come as no surprise that Alfred Packer, the man-eater of Colorado, has had a rabid following of persons lured to the daunting mysteries of his strange case. Some have even tried to vindicate his name from the nastiness that has surrounded it for over a century. Some of those apologists for "Alfie" have sought to convey their mordant views on the stage, as in *Cannibal! The Musical,* or in song, as in "The Ballad of Alferd Packer," performed by The Alferd Packer String Band of Lawrence, Kansas. Those renditions of the cannibalism of Alfred Packer ill-conceal the loathsomeness of cannibalism, especially when preceded by murder.

How I came into this case, and made it my first exhumation, was by chance and serendipity, but it radically changed the focus of my work in forensic science from the classroom to the field.

I had known about Alfred Packer from a footnoted reference in the casebook that I used to teach the basic course in criminal law at The George Washington University. That footnote had left me with many twinges of unfulfilled curiosity about the whys and the wherefores of the saga of Alfred Packer. The spare outlines of his story were well-known, albeit hotly disputed.

In the beginning, Packer was an ordinary man, save for his inability to decide whether he was to be called Alfred or Alferd. Until he emerged from the winter vastness of Colorado's San Juan Mountains in April of 1874, his accomplishments had not risen above the mediocre. But from that time, his life took on a new and very public meaning. Not only had he conquered the harsh winter, the mountains, and his own recurrent bouts of epilepsy, but from all appearances he had done it alone. Or had he? His tenure as a perceived miracle worker was of short duration, and people soon realized that he had done the unthinkable.

IN 1873, THIRTY-ONE-YEAR-OLD ALFRED PACKER, IN THE COM-pany of a group of twenty-one other prospectors, left the safety and security of their lodgings near Provo, Utah, and headed southeast on the Mormon Trail, looking for better mining prospects in Colorado. The party arrived in January 1874 at Chief Ouray's Ute Indian camp in northwestern Colorado, where they were warned by the Indian chief to remain until spring brought weather conditions more favorable for their travel. At that time of year, the mountain passes through the San Juans that were within their sight east of present-day Delta were treacherous and all but impassable, so they were reliably told.

Nevertheless, six of the prospectors, including Alfred Packer, apparently frenzied by the prospecting spirit, were determined to continue without delay. On foot, with provisions sufficient only for a short and quick trip and with only one rifle among them to enable them to forage for game, the doomed men left Chief Ouray's camp on the ninth of February 1874. In the group were Alfred Packer, Shannon Wilson Bell, Israel Swan (the oldest among them, at fifty-some years of age), James D. Humphrey (or Humphreys), Frank "Reddy" Miller, and George "California" Noon, who at eighteen years of age was by far the youngest of the group.

Only Packer completed the torturous journey. The only evidence of what happened to the others was contained in his multiple and conflicting accounts of the trek and its numbing hazards. In pressing east along the Gunnison River toward what is now known as the "Black Canyon of the Gunnison," they somehow lost their bearings and veered south on the Lake Fork of the Gunnison toward its terminus in Lake San Cristobel, near present-day Lake City, Colorado. Heavy snow and bitter cold hindered their progress and eventually brought them to a standstill, with their provisions exhausted.

The streams and lakes were too frozen or treacherous to fish, and wild game was nonexistent. They turned to rose hips and to sucking on their shoe leather for meager sustenance as the snow piled high around them. Their situation was as bleak as the snow-filled skies under which they were trapped.

Packer arrived alone at the Los Piños Indian Agency near Saguache, Colorado, on April 16, 1874. His statement to "General" Charles Adams at the Los Piños Agency indicated that the other five men had died at various stages of their journey, either as starvation overtook them or when they were killed in self-defense from one another's attack.

Then he said that there had been a fierce snowstorm and the party, one by one, had been lost along the way. Swan, being the oldest at around fifty, had died first, and they had taken pieces from him to eat. Then, four or five days later, James Humphrey died and "was also eaten." He had $133 in his pocket, and Packer had taken it. The third man to die was Frank Miller, an "accident" that happened while Packer was searching for wood, and the next to go was the boy, George Noon.

Packer reported that while he was off for several days hunting for game, Bell had shot Noon with Swan's gun. Packer had come back, and together they ate him. That left only Packer and Bell. "Bell wanted to kill me," Packer's May 8 report indicates, "struck at me with his rifle, struck a tree and broke his gun."

In another of his signed statements, given many years later, Packer stated that a crazed Bell killed the other four with a hatchet while Packer was absent. Upon Packer's returning to their temporary camp after foraging for food, he said he found Bell sitting at a fire roasting a large piece of meat, "which he had cut from the leg of the German butcher" (apparently referring to Frank "Reddy" Miller). The other three men were said to be spread out in death near the fire, their foreheads split open by hatchet blows.

Bell then attacked him, so Packer insisted, forcing him to shoot the man from the side in the belly in self-defense. When Bell dropped his hatchet as he collapsed to the ground, Packer picked it up and finished him off with a couple of blows to his skull. Packer was taking no chances that Bell would survive the bullet he had put in him.

He tried to leave the camp, but the snow was too deep and forbidding, so he was reduced to living off the flesh of his dead fellow prospectors until the spring thaw enabled him to escape the site, taking with him pieces of their flesh to consume along his escape route.

At some point, probably at his homicide trial, Packer is said to have drawn a sketch of his dead companions laid out around the fire pit. To me, having seen the sketch, it looks less like a carefully crafted map of the location of the bodies than an outline preliminary to a Rorschach test.

Oddly, when Packer came in from the mountains, he was well-heeled with cash and possessions, whereas he was known to have been without funds when he and the other twenty-one would-be prospectors left Utah. To add further ominous suspicions to his reportage, he professed having gone for more than a day without food, but in spite of his asserted hunger he asked for nothing to eat.

Suspicions of his treachery mounted when he was seen to be in possession of Noon's Winchester rifle and when an Indian guide found strips of human flesh along Packer's flight route. (How the flesh was determined to be human is nowhere explained.) Those who had heard his constantly changing tales surmised that his recounting of the occurrence was far different from what had actually happened. Perhaps in the group's dire plight he had killed the others, motivated by a consuming desire for their money and other possessions. After all, they were all likely to die anyway, what with the weather and their weakened physical condition and their lack of provisions. Packer might have reasoned that at least he would die in possession of valuables such as he had not lawfully acquired during his lifetime.

A search party was organized and sent out, led by a reluctant Packer. But it failed to find the missing prospectors, it later appearing that he had conscientiously misdirected it from the site where the bodies were ultimately discovered quite accidentally by a vacationing artist, John A. Randolph.

It was in August 1874 that Randolph, an artist with *Harper's Weekly* magazine, came across the skeletal remains of five people in a grouping adjacent to the banks of the Lake Fork of the Gunnison River just two miles from the present-day Lake City. Among the remains were pieces of torn clothing, blankets, and some shreds of flesh, but the weather and the actions of carnivores had plainly pillaged the bodies. Randolph found no shoes, cooking utensils, or guns among them. It appeared that they had not only been killed but also horribly ravaged, with one body missing its head. Randolph reported his grisly discovery while pausing in cool detachment to sketch them as they lay in unperturbable death. As published in *Harper's Weekly,* his sketch does not comport in many respects with the factual and scientific realities of the event. California Noon, the teenager, is shown with a full adult beard and an equally adult severely receding hairline that would not likely be seen on a person of his youth. Noon is also inexplicably portrayed with only a short segment of his left humerus (upper arm bone) attached to his shoulder girdle, while both Noon and Bell, who was positioned in the middle of the group of five, are drawn with fully defleshed skulls. Scientifically, the feeding of insects lighting on the bodies would have been the first and the foremost evidence of fleshy decay, particularly in and about the orifices of the head, not to mention the hatchet-lacerated tissue of the head of Shannon Bell.

The accuracy of Randolph's sketch was put further in doubt by the fact that each one of the five persons depicted by him is given a name below it, the names corresponding to the five missing prospectors. But were the names truly those of the persons drawn by Randolph? There is no documentation as to who it was who identified the deceased persons. Certainly Randolph could not do so on his own. It awaited my exhumation in 1989 to reveal the truth.

Randolph's discovery brought officialdom to the alert, resulting in Packer's being arrested and jailed in Saguache. Meanwhile, the Hinsdale County coroner, W. F. Ryan, visited the death scene in the company of twenty other people, purportedly to hold an inquest, but no written documentation of it has survived. One member of the original party of twenty, Preston Nutter, identified the remains. The headless corpse was said to be that of Frank "Reddy" Miller. No explanation for how these identifications of the skeletal remains were made has been preserved for posterity. The remains of the five were then buried, according to a second Randolph sketch, cheek by jowl with the adjacent Lake Fork of the Gunnison water course. A third sketch was executed by Randolph, enigmatically visualizing a lateral view of a short but steep incline. The significance of this drawing was a mystery until my exhumation in 1989 shed light on it. For years after Randolph's sketching, continuing even until 1989, the site depicted in Randolph's drawings was known as "Dead Man's Gulch." It might just as well have been termed the "Golgotha of the Lake Ford of the Gunnison."

After the burial party finished their noisome task, it was learned that Packer had escaped from the sievelike Saguache jail. He managed to elude the law for nine years by living under the assumed name of John Schwartze. It wasn't until March 1883 in Fort Fetterman, Wyoming, that French Cabizon, a member of the original party of prospectors, chanced upon him some three hundred miles from the scene of the crime. Unmasked and rearrested, Packer was presented to a Hinsdale County grand jury, which returned five indictments against him for the hatchet murders of the five hapless prospectors.

Packer was tried in the Hinsdale County courthouse in Lake City, Colorado, separately for the murder of Israel Swan, the one whose age made him the easiest to dispatch. Georgia native Melville B. Gerry presided. Preston Nutter, a member of the original group of prospectors, who had identified the five victims for the coroner, testified as a key witness to what he knew. He described for the jury the positions of the bodies as they were found, pointing out the hatchet wounds to the heads, which he had personally observed. One of the five was described as having been struck hard in the back of the head, causing it to be crushed.

Oddly, the coroner, the man presumably in the best position to offer a professional opinion, did not testify. Since he had never penned his observations on the condition of the remains, there was nothing from him to guide the court. In fact, no one experienced in the science of criminal investigation testified at Packer's Lake City trial.

When recalled later in the trial, Preston Nutter described a hole in a bone detached from one of the bodies that, in his layman's opinion, was said to have the appearance of a gunshot wound. He also described the clothing as having been cut and ripped up.

Taking the witness stand in his own behalf, Packer defended himself in an uninterrupted statement, as then allowed in Colorado, for over two hours, and in the process told a number of significant lies. He lied about his age, his military service for the Grand Army of the Republic during the Civil War, or as southerners describe it the War Between the States, and the cause of his epilepsy, which he attributed to his strenuous exertions in walking guard duty. In fact, as revealed by the military records for the Union forces now housed at the National Archives, he had enlisted twice in the military. The first time, he received a medical discharge due to his being afflicted with epilepsy, and then, enlistment bounty hunter that he was, he re-upped. Ten months later, he received a second medical discharge.

Addressing the killing of Swan, he professed to be innocent of it, but he did admit to having shot and killed a hatchet-wielding Wilson Bell in self-defense. He also spoke of the deaths of the others, and said that some of those who had survived longer had eaten the others to stay alive. However, when all of this gruesome activity allegedly had occurred, Packer himself had been scouting for an escape route for food. He returned to find human remains already boiling in a stewpot, although he did admit to taking meat from the bones of two of the deceased (Bell and Miller) to slake his own hunger. Since he'd offered several different versions of his experiences at different times, and had confessed to taking the victims' personal belongings, and probably on account of his uncontrolled loquacity, he was perceived in the worst light as truculent, overbearing, and flippant.

Although Packer might have thought he had carried the day when he made his case with his detailed and rambling presentation, the jury was not taken in and cold-shouldered his self-defense claim. On Friday, the thirteenth of April in 1883, Alfred Packer was convicted of premeditated murder, with Judge Gerry sentencing him to be hanged in a declaration that has resonated through the years as the capstone and the hallmark of the Packer case—surely akin to the popular singsong refrain that later captured the public's fancy in the killing of Lizzie Borden's father and stepmother.

Judge Gerry is said to have momentarily cast off his polite and proper Georgia upbringing to sentence Packer in these ringing and memorable words:

"Stand up, yah man-eatin' son of a bitch, and receive your sintince . . . They was siven Dimmycrats in Hinsdale County, but you, yah voracious man-eatin' son of a bitch, yah et five of thim! I sintince ye t'be hanged by th' neck until y're dead, dead, dead; as a warnin' ag'in' reducin' the Dimmycratic popalashun of th' State."

Other versions of this semiliterate, ungrammatical, and wild political diatribe qua sentence abound. However, the tradition that has evolved around Judge Gerry's supposed sentence is a myth, perpetrated by an Irish publican from Saguache, Colorado, the town where Packer had been temporarily imprisoned. In fact, Judge Gerry's actual sentence was a model of judicial clarity and literary craftsmanship. In part it said:

"You, Alfred Packer, sowed the wind; you must now reap the whirlwind. Society cannot forgive you for the crime you have committed. It enforces the old Mosaic Law of a life for a life, and your life must be taken as the penalty of your crime. I am but the instrument of society to impose the punishment which the law provides. Where society cannot forgive, it will forget. As the days come and go, the story of your crimes will fade from the memory of men. . . .

"Close up your ears to the blandishments of hope. Listen not to its flattering promises of life; but prepare for the dread certainty of death. Prepare to meet thy God; prepare to meet the spirits of thy murdered victims; prepare to meet that aged father and mother of whom you have spoken and who still love you as their dear boy."

The unmasking of the mythical nature of Judge Gerry's fictitious but popularized sentencing of Packer is not done without a certain trepidation, since the myth does make a frightfully good story. But it was only the first of many myths surrounding Alfred Packer that my investigations in the ground would expose.

However, two years later Packer won a new trial, to take place in Gunnison, about thirty miles away. The Colorado Supreme Court set aside the murder conviction based on a legislative oversight, which rendered ineffectual the many official invitations to Packer's hanging. Packer could not be tried in 1883 in the state of Colorado for a murder he had committed in 1874, while Colorado was still a territory, statehood having occurred on August 1, 1876. In enacting the criminal code of the state of Colorado, superseding that of the territory, somehow the legislature had failed to grandfather into the new code murders committed during the territorial days. So

Packer was retried for all five incidents on a different charge: manslaughter as it had been carried over to the new code, permitting Packer's being re-indicted for the manslaughter of his five traveling companions.

The venue for the second trial was in the more populated Gunnison. Once again Packer was convicted, and this time he was sentenced to five consecutive sentences with a maximum eight years each, totaling forty years in all.

Neither on Packer's first, Lake City, murder trial nor on his second, Gunnison, manslaughter trial was he ever charged or convicted of the crime of cannibalism. Despite the folklore, widely publicized, that Packer was convicted of cannibalism and, indeed, that he was the first person in this country ever convicted of that unspeakable crime, the facts put the lie to the myth.

First, cannibalism was not a crime in the Colorado or Territorial statute books either when Packer's crimes were committed or when he was twice tried. Furthermore, cannibalism would not likely be the choice *du jour* for a prosecution of Packer when five victims of homicide deserved vindication.

Even in today's world of criminal statutes in the fifty states, the only state declaring cannibalism to be a punishable criminal offense is Idaho, where a maximum fourteen-year term in the state prison is prescribed upon conviction for that crime. The Idaho statute defines cannibalism as "willfully ingest[ing] the flesh or blood of a human being." A complete defense is engrafted onto the statute where "the action was taken under extreme life-threatening conditions as the only apparent means of survival."

That proviso might have pleased Packer, if applicable in Colorado, always supposing his admitted cannibalistic feasting of the bodies of his prospector buddies would constitute taking "the flesh or blood of a human being." The issue for legal haranguing, at least in Idaho, would be whether cannibalizing a dead body is cannibalizing "a human being."

After serving sixteen years in the Canyon City, Colorado, penitentiary, Packer was paroled—not pardoned, as some have written—in 1901 (another of the myths that abound), and on April 24, 1907, he died from a stroke (listed on his death certificate as caused by "senility—trouble & worry"). Buried under a Grand Army of the Republic souvenir–collector-chipped headstone in a cemetery in Littleton, Colorado, Packer has many supporters who firmly believe that he was a victim, not an offender. As one view has it, his acts were excusable as having happened *in extremis*. Another position exonerates him entirely as having killed only one man, Shannon Bell, and then only as a desperate man acting in self-defense.

Packer's military service gave him a grave at government expense. It also gave him a pension owing to his disability discharges. His disability was epilepsy contracted, so it was alleged in his petition for a pension, while walking guard duty. This assertion was absurd on its face and even laughable when it is realized that the affidavits supporting his claim to a pension were signed only by a number of his fellow inmates at the Canyon City Penitentiary. What these facts paint in glowing colors is a man whose life was marked by lies and deception. As a result, the government was swindled out of a monthly pension payment to Packer. That doesn't make him a murderer, but it does color any effort in support of his exoneration for having given credible explanations for his actions at "Dead Man's Gulch."

Years later, in 1928, the citizens of Lake City erected a monument to the five victims at the top of an incline within sight and sound of the rushing waters of the Lake Fork of the Gunnison a mile or so outside of Lake City. As part of the monument-raising ceremonies, the community threw a fish fry at the site. But whether the stone marker erected in 1928 was a monument only to the memory of the victims or was meant to be a headstone for their grave site was cloaked in uncertainty until my probing of the site in 1989.

UNTIL 1989 THE SAGA OF ALFRED PACKER WAS JUST AN UNSETTLED story to me, but that year I was to have the first opportunity, using scientific knowledge and instrumentation, to discover once and for all the truth of this garish and fiction-riddled tale. Thus far, the tale of Alfred Packer had only been the plaything of historians. I would be the first to bring the clues and advantages of forensic science to bear upon it. How I got involved in 1989 was almost as quirky as the Packer story itself.

While teaching my criminal law classes over the years on the subject of the defense of necessity against charges of homicide, I would point out that the most remarkable documented instances were those involving cannibalism in extreme situations of deprivation due to starvation. The most celebrated situation of this kind was spelled out in the British case of the 1884 trial in England reported as *Regina v. Dudley and Stephens.*

It seems that a sailing ship sank on the high seas with the survivors having crowded into a lifeboat. As matters went from hunger to starvation, the youngest of the group was, without even the pretense of any attempt at a fair selection process, killed and cannibalized to save the others. Their defense at their subsequent murder trial was bottomed on the necessity of the circumstances. In the casebook report of the holding in *Regina v. Dudley and Stephens,* there is a footnote reference to the Packer case, which up to that time was the leading United States case involving cannibalism. (Milwaukee's Jeffrey Dahmer has bested Packer in his man-eating proclivities.) It troubled me that Packer had claimed only that he had killed in self-defense rather than asserting the necessity of the treacherous surrounding circumstances as a justification for his acts after the example in *Dudley and Stephens.*

Then, during the summer of 1988, when I was a spectator at a meeting of the board of directors select committee of the American Academy of Forensic Sciences in Colorado Springs, the board quite fortuitously provided me with the opportunity to look more closely into the many legends of Alfred Packer. My reputation in the academy was as an enfant terrible (even in my middle-age years), always on the prowl like "the wolf who keeps the caribou strong" by thinning the herd, as Dr. John Thornton once referred to me in introducing me as a speaker at an Academy of Forensic Sciences function.

I went to the first day of the board's meeting, and at the end of the session I felt the unpleasant and undesired sting of ostracism settling upon me. My imagination was not working overtime. On the second day, I decided against spending another fruitless and enervating day being a bump on the board's agenda. So when I found the room where the board had met the previous day vacated and locked, I beat a fast, somewhat paranoia-generated, track out of Colorado Springs with my sights set on Lake City and my mind turned from the dry bones of the board's meeting to the aging bones of Alfred Packer's five prospector companions.

What better way for an investigative forensic scientist to conduct his research than to revisit the scene of the crime and the burial place of the victims. Lake City is definitely an out-of-the-way place. No airport services it, and at an elevation of better than 8,000 feet it is remote and untrafficked and underpopulated, some say with fewer than three hundred year-round residents.

As I sped (or was it fled?) westward out of Colorado Springs, climbing the long hill with panoramic vistas to the east, I had little time or reason to dwell on the board of directors meeting I had left behind. My sights were set on an historical adventure that would take me I knew not where.

Alone in my rental car, I traversed a wondrous landscape, becoming less well traveled the farther my wheels took me to the remoteness of Lake City. The canyons with their roads following the adjoining water courses were asplash with sun-glistening dancing light. The summit at Monarch Pass was the occasion for a picture-taking stop and rubbing of the hands to keep out the chill of the snow-speckled mountain promontory.

The road leading into Lake City is narrow, winding, and, for the most part, downhill. A sign on the roadside indicates the presence of a "massacre site." Passing it without connecting it to Alfred Packer, as I should have, I continued the short jaunt into town, where the main street was littered with shops selling diverse wares attractive to tourists, including any number of Alfred Packer T-shirts. One such portrays a grizzled-looking Alfred Packer holding a fully fleshed bone to his mouth while urging all viewers to "Have a Friend for Dinner." Most assuredly, I had arrived in the motherland of Packer legend and lore.

I bought a tacky T-shirt while asking for directions to the notorious Cannibal Plateau, reputed to be the burial place of the five deceased prospectors. I was directed to the high hills above the town where, I was assured, there was a prominent plaque marking the place as Cannibal Plateau. With these loose travel directions in mind, I promptly left Lake City in the hope that Kurt Vonnegut would be proved to be right, that these "peculiar travel directions would turn out to be dancing lessons from God."

As it turned out, the hills I searched on foot were marked by stands of glowing aspens, but no sign of Cannibal Plateau stood out to welcome me. Either this professorial inquiring tourist had been deliberately misled or the sign had been removed to fill up the wall space in some college student's dormitory room.

Unrewarded, I returned in a griping and glum mood to the commercial downtown in Lake City. Noting a building housing the offices of the *Silver World,* the local newspaper, I stopped there hoping to minimize my so-far fool's errand with more precise and accurate details. It was then that my good fortune began to shine on me.

I met with Grant Houston, the editor of the *Silver World,* the town's historian, and a man with as quick and glowing a mind as the stream where he took me to share a lunch. His interest in and knowledge of the many Packer legends was as clear and capacious as the blue skies above us.

"Yes," he said, "the dead prospectors were all buried in the vicinity of Lake City in spite of a large body of opinion insisting even to this day that they had been interred miles away on the shores of Lake San Cristobel, the terminus of the Lake Fork of the Gunnison River."

The only "Cannibal Plateau" to his knowledge was the elevated and flat area just above a short incline from the Lake Fork on the outskirts of Lake City. The city folk, according to him, had dubbed the place where the prospectors died as "Dead Man's Gulch."

I was admonished that the original Dead Man's Gulch was no more, the Lake Fork having overflowed its banks in rechanneling over the years, leaving the Gulch and all its grisly reminders under water. My hopes of finding the prospectors' burial place were fast fading. I was on the verge of calling off my search.

Grant offered to take me to the site that he firmly believed was the true burial place of the deceased prospectors. I leaped at the opportunity. So back in our cars we went, reversing my original route into Lake City. We traveled only as far as the enigmatic sign noting the presence of a massacre site somewhere in its vicinity. We left the main road and took a side road to the left until Grant pulled up next to a mammoth boulder enclosed by wooden fence posts. Affixed to the boulder was a metal nameplate containing the names of the five prospectors who had not survived the winter of 1874 when traveling with Alfred Packer.

The moment was one for restrained ecstasy on my part, restrained because a shadow had crossed the expressive face of Grant Houston. "This stone," he said, "was placed here, many say, as a memorial to the deceased prospectors, not to mark this as the place of their burial. In your business I am sure you are familiar with such places, such as many of the memorials in Washington, D.C., where the persons memorialized have been buried elsewhere."

Cutting to the chase, I asked, "Are you saying that this boulder does not represent the place of final repose of Packer's five companions?"

"Not at all," he replied. "Let me show you something up close to the boulder."

And lo and behold, he pointed to five smaller boulders, like stepping-stones spaced a few feet apart and running from one side of the boulder to the other. They were almost indistinguishable from the mammoth boulder arching above, the effluvium of soil runoff having left only their tops looking out to perplex the onlooker.

"It is my opinion," Grant continued, "that those are the markers signifying that this and no other place is the authentic burial place."

The number and alignment of the stones tempted me to agree, but, long-term and well-schooled skeptic that I am, I wanted further corroboration.

Grant perceived my uncertainty and pointed to a level area as likely to be the Cannibal Plateau that I had failed to find thus far. To support that viewpoint, he walked me across the flattop area to a point where the rechanneled Lake Fork lay coursing its way southward below us. Down this incline to the river lay Dead Man's Gulch, where a sketch was made of the five deceased prospectors by John Randolph. "In Randolph's three sketches," Grant continued, "we see Dead Man's Gulch with the prospectors arrayed as if in readiness for burial. In another sketch, Randolph depicts a lateral view of a treeless incline, which is suggestive of this very incline and also intimates the likelihood that the bodies of the deceased prospectors were carried to the plateau at the top of the incline as portrayed."

"The final pictorial found on the front cover of *Harper's Weekly*," I chimed in, "could well be that of the bodies of the five prospectors readied for a quick and simple and coffinless burial."

Grant was making his case for this site as the only burial place, and he was doing it with compelling detail and lawyerlike persuasiveness. "But I see you are still not sure that you should commit to my viewpoint that this is the one and only grave site for those hapless, murdered prospectors." He had intuited my hesitation over this highly nonspecific, circumstantial evidence almost as if he were able to probe my mind. "There is one more piece to this puzzle of identifying the location of the grave site which I have held back," said Grant, as if he were layering me, à la Agatha Christie, with one fact after another, leaving me with no alternative but to concur with his ultimate conclusion. "We have a seasoned and expert photographer in town who can show you prior pictures of this site, even some taken many years before the presence of this boulder and these fence posts. Those pictures, which go back a hundred or more years, unequivocally demonstrate that this and no other place is the burial place."

I was most attentive to this new information, but it was not until later that I had the occasion to view the pictures to which he had referred and to compare them to topographic features. All the pictures, with similar background mountains and trees, suggested this site as the most likely burial place. Utilizing photogrammetry had been most rewarding.

My satisfaction, at least presumptively, at having found the elusive burial place left me with a new concern. If an exhumation were to be conducted, who was the owner of the property, and would the owner consent to an exhumation on the land along with all the subsequent scientific work that would be necessitated? Once again, Grant came to my rescue: "The property on which we are standing is owned by an Oklahoma neurosurgeon and his wife, Joe and Sandra Jarman."

That news spurred me to see if the owners were at home in their house, lying just a hen's kick from the boulder-monument. My knock was answered most warmly by Sandra Jarman, who, after I identified myself and explained my mission, invited me to join her and her husband in their house for iced tea. Our conversation went swimmingly.

I explained to them that I was investigating the many tales of Alfred Packer and the death of his fellow prospectors. Courteous and intrigued, they asked me how I thought I could add new details to the historical record of Alfred Packer. I proposed that modern forensic science could play a role in confirming or discounting Alfred Packer's multiple explanations.

"And how do you think you can accomplish that?" they asked simultaneously.

"By an exhumation of the remains of the prospectors buried on your property," I audaciously proposed.

It was a radical idea, and not something I'd ever done before, but being a man of science himself, Dr. Jarman recognized that my proposal might have merit and was quick and unconditional in his support of it. Permission to exhume the bodies on his property from him and his wife was immediately forthcoming and unequivocal. At that point, I took them at their word. In later exhumations, however, I have learned to be more lawyerly in drafting written consents both for the exhumation and for the scientific testing that follows the exhumation. There is no form for such written consents, for each exhumation bears its own marks and perplexities, causing the consents to be tailored to the exhumation.

That encounter with Grant and the Jarmans was the unexpected and unplanned invitation to what was to become a lifetime enterprise. I set as my objectives the twin interests of science and the voiceless victims. If Packer had murdered them, I wanted science to prove that fact, if only to give rest to the remains of his victims. Thus far, public and historical attention had been paid to Packer almost exclusively and little to the deceased prospectors. I wanted to right this historical and public imbalance. There were two full-length books, a play, a movie, and several ballads devoted to Packer, some of which had been sung by The Alferd Packer String Band of Lawrence, Kansas. There had even been a determined movement to place a bust of him in the Colorado state capitol together with a posthumous pardon from the Colorado governor. One author said that Packer had taken on a heroic stature. Fortunately, no one has put him on a par with a Siegfried of Wagner's Ring Cycle or as a Ulysses of the Homer epic.

Few people even knew the names of the prospector victims. I wanted to give them their day in court, which through science I was convinced I could do. Since the remains had been buried for over a century, time was clearly of the essence, lest the ravages of time savage my hopes.

Bones can tell many tales when conscientiously analyzed by a careful and experienced physical anthropologist, at least if they are in adequate condition for that purpose. They can bear witness to bullet holes, to bodily trauma caused by blunt-force impact inflicted by objects like hatchets, rocks, rifle butts, hammers, or by striated objects like a hatchet or knife blade. Bones will also show the tale of the scrape marks from any deliberate cutting of them. They can certainly signal knife gouges or cuts indicative of defleshing in connection with cannibalism, or in the disarticulations that occur in an effort to sever the remains. In doing so, the true cause and manner of the death can be masked or the identity of the deceased obscured so as to frustrate law enforcement and to distance the perpetrator from the killing. I was more than eager to subject the bones of these prospectors to a close forensic anthropological analysis.

MY FIRST TASK, HOWEVER, IN DETERMINING THE VALUE OF AN exhumation was to take soil samples at the burial place, because the pH of soil and its mineralogy dramatically affect the rate of deterioration of human remains. Indeed, the subject of the postmortem fate of human remains in various environments has developed into a new forensic discipline styled *forensic taphonomy.*

The site of the boulder-monument was relatively treeless, which meant little damage from tree roots seeking nourishment or just growing through the buried remains. The drainage of water appeared to be good and the burial place was on the sunny side of the mountain, indicating that snow would not gather for long at the site. The soil, when tested, proved to be pH neutral, in the range of 6 or 7, as well as sufficiently permeable that water would not be retained in the soil, causing the bones to rot with decay. The wooden fence posts surrounding the boulder had been in place for sixty years, and probing the wood underground revealed them to be undecayed and intact, a promising sign for the preservation of the prospectors' remains.

Along with the archival pictures of the site provided by the Colorado Historical Society and the Colorado State Archives, the information I had in hand all pointed in the direction of a viable exhumation project, but as

I am constantly reminded by the literature and my scientific colleagues, you never know until you are there. And then, of course, it is too late to bemoan your fate. Pre-exhumation planning to the nth degree is thus a categorical imperative.

I didn't know how the bodies had been prepared for burial, but it was unlikely they had been buried in coffins, considering the putrefied and at least partially skeletonized condition of them when discovered. I learned from records archived in a vault in Lake City's courthouse that there had been an undertaker's charge of $37 paid by the city, meaning that the victims had had some semblance of a proper burial. What I did not know was how deeply the burial pit had been dug or whether the five sets of remains had been buried separately or commingled after the fashion of an ossuary (a container for the bones of many persons). Nevertheless, armed with the results of my soil analysis, I was ready to move ahead, especially since the photogrammetry analysis supported my own view that the Jarman property was the actual burial site.

Over the next several months, I organized a team of forensic experts to contribute their specialized expertise to the exhumation of the remains of the five men from Packer's party. When the press got wind of my project, a front-page report from the *Silver World* of Lake City started a media barrage, and considerable interest was aroused. Our chief forensic anthropologist, Dr. Walter H. Birkby, Curator of Physical Anthropology at Arizona State Museum in Tucson, invited his colleague, James E. Ayres, a professional archaeologist and consultant to the National Park Service, to participate. For a thorough post-exhumation analysis, we were granted permission to use the forensic anthropology lab at the University of Arizona in Tucson, but that would mean transporting the remains across the Colorado state line. As a consequence, I was confronted with a not-insubstantial legal question: Were there any laws governing my transporting the remains across a state line? That question was submitted to my lawyer team members, especially John "Jack" Carraher, a lawyer with the Denver law firm of Feder Morris Tamblyn & Goldstein, who advised that I had no worries or paperwork to complete in that regard.

Walt also had two doctoral students enthusiastic about taking part, and I had the consulting services of my friend and colleague, Dr. Douglas H.

Ubelaker, the Curator of Physical Anthropology at the National Museum of Natural History at the Smithsonian and frequent consultant to the FBI. Dr. Allen Jones, the deputy chief medical examiner from Tucson, and Dr. Bill Anderson, a volunteer forensic pathologist from Florida, were prepared to consult on tissue samples should we find any. In addition to them, we had an engineer, prophetically named Packer; two firearms examiners, George R. Wilson and Anthony (Larry) Paul; two document examiners, Doug Cayton and Duayne Dillon, to verify the authenticity of Packer's two signed confessions; and various and sundry experts in geophysics, tool mark identification, research, and backhoe operation. We listed Lake City local Byrne Smith as our forensic backhoe operator. That characterization caused many a smile of amusement. And why not? After all, cannibalism leaves little room for humor, unless it be mordant.

I also had a photographer, Jim Kendrick, from the medical school at the George Washington University; my personal research assistant, Sherrie Hardwick, who helped to get things organized; and a journalist in applied science, Patrick Zickler, who was also our coordinator and fund-raiser, and who brought his four-year-old daughter. The final person to round out the team was my wife, a most knowledgeable amateur historian.

I searched for a project name or title that would provide a theme for this venture, but I failed to find one that fit. I gave the subject ample, somewhat distracting thought as I bicycled home the fifteen miles from my university office. One day, upon opening the door to my home, my three-year-old grandson, Willie, threw his hand in the air and said, "Gimme me five, Grandpa!" He'd obviously come upon a new expression, signaling enthusiasm and promise. "That's it!" I said. "Gimme five." I wanted five skulls and the five bodies to go with those skulls. That was the alpha and the omega of the project.

Consequently, "Gimme Five" became the team's ringing theme. To firm up the point, I crafted a T-shirt emblazoned with that slogan along with the skeletal outline of the human anatomy depicted by the father of anatomy, Belgian Andreas Vesalius. As a reward, three-year-old Willie became a full participating member of the exhumation team, and the date for the commencement of the exhumation was set for July 17, 1989, Willie's fourth birthday, with Willie scheduled to be present at the site.

Having learned that we did not need permits from the state board of health and that Colorado did not have statutes requiring a lot of legal work preliminary to an exhumation from private property, I nevertheless thought it prudent to have Jack Carraher present to keep everything on the legal up-and-up.

Just in case there might be some living descendants of the deceased prospectors, I put a notice in the *Silver World* of Lake City publicizing to the world, or such of the world who might have access to that local newspaper, my planned exhumation. There was no statutory requirement for such a notice, but I was conscious of the need to play it safe. The lawyer in me was speaking.

The Jarmans did impose a requirement that I secure a liability insurance policy protecting them against any possibility of legal action. The insurance cost, there being no actuarial tables for comparable projects, was a whopping $6,000 per month. This hefty fee imposed a one-month limitation from exhumation to reburial as the project's timetable. I had no funding, except personal resources, to pay for such liability insurance beyond one month.

Although I had a team made up of volunteers, the expenses mounted. We figured it would cost roughly $12,000 to $15,000 for the entire project. We were therefore on the constant lookout for funding to carry the project through. A local resident donated $1,000 to Scientific Sleuthing, Inc., a tax-exempt, nonprofit corporation that publishes the *Scientific Sleuthing Review,* which was used to defray some of the costs we would incur. The sale of T-shirts and project manuals brought only a minimal return. Other than small acorn grants from the American Academy of Forensic Science and The George Washington University, along with a few private donations, I personally bore the brunt of most of the costs.

Not being independently wealthy and having no annuity to support me, I gave considerable thought to the risks of failure confronting me. I was, however, no stranger to risk-taking, having bicycled across the breadth of the United States on three occasions. I had also experienced the hazards of being a civil rights lawyer in Mississippi in 1965 and 1966. There were many and varied risks in the Packer project, but the prospect of not succeeding never really caused me any grief.

Naturally, I had my share of detractors and unwell-wishers. Some thought what I was doing was macabre or grotesque, while others believed I was wasting my time. The majority of my naysayers voiced the same refrain: "You're going to dig down and you're not going to find anything. There won't be any bones." Even if there were bones, some of them said, they'd be commingled and impossible to separate. If they weren't, then they'd be beyond our means to make any scientific determinations. There were even those who insisted with a nineteenth-century author named Jocknick that the burial place was on the banks of Lake San Cristobel many miles from the Jarmans' property. Being of stubborn Irish heritage and having done my preliminary scientific homework, I was not to be dissuaded.

Not even Doug Ubelaker, a renowned specialist in the field, could deter me—even when, at the Smithsonian Institution, which he calls home, he showed me some bones of a similar age that had been dug up in Pennsylvania. When he put the bones in my hands, they just crumbled like bone meal. "That's what you're going to find," he said confidently.

But I stood my ground. In spite of his expertise as a forensic anthropologist, I had the firsthand data gleaned from the burial place in Colorado and it all argued for well-preserved remains. I was bold enough to reply that I had done the necessary preliminary workup of soil analysis and water runoff, which convinced me that the bones would be intact and in analyzable condition. And that's where we left it, he going off to foreign lands for one of his many digs and I heading off to Colorado for my first such dig.

Admittedly, my agenda was somewhat different from that of the scientists working with me (or better, with whom I was learning). They were there not because they had an eye to success but because either they possessed the insatiable curiosity of true scientists or because this was such an engaging investigation. My purpose, if not also theirs, was to clarify the historical record, particularly when confronted by the many larger-than-life myths surrounding Alfred Packer.

My research had also turned up an aboveground issue that had taken on the character and dimensions of a myth. Whenever Packer's name appeared on legal documents, the signature was that of Alfred. But there were those who pointed to a tattoo on his arm that read "Alferd" and to the fact that nonlegal documents were signed by him with the name "Alferd." My document examiners perused the relevant documents I supplied to them and decided that Packer had sometimes signed his name as "Alferd" and sometimes as "Alfred." I thought it best and lawyerlike to stay with the "Alfred" designation, since the legal documents he signed bore that name.

As I prepared for the dig, Grant Houston and I shook hands on a gentleman's bet regarding the number of skulls I would find buried in the grave. According to historians, when the remains were initially discovered by Randolph, and as sketched by him, only four skulls had been seen at the site. It was Frank "Reddy" Miller who was said to be the headless one.

A year or so after the discovery and burial of the bodies, as the story has it, a skull had been located some distance from the site. My confidence in the project's success knowing no bounds, I was willing to wager that we would find five skulls. After all, hadn't Willie, my grandson, christened the project as the "Gimme Five" investigation?

"You're only going to find four," Grant insisted.

"How do you know?" I asked.

"Because the tradition is that one of the skulls was found and put on the desk of a local bank president. It even had hatchet marks on it. When you got a loan, he'd turn it to the right, and when you didn't, he'd turn it to the left."

This fanciful tale of an urban (mountain) legend nature had many variations among Packer commentators. Some said that Frank Miller had been found without a skull. Yet the Randolph illustration seemed to show that two heads were missing, those of both Israel Swan and Frank Miller.

Since the scene had been rendered for the purposes of a magazine, the artist was not to be altogether trusted. Certainly, eighteen-year-old George "California" Noon would not have had the receding hairline that Randolph gave him. Nor would Noon and Bell have had the fully fleshed faces as depicted by Randolph. Insects would have found a home in the orifices of the head, establishing their feeding colonies there soon after death. The skulls should have been found denuded of flesh, as James Humphrey's was in the Randolph sketch.

I knew that in his courtroom testimony, Preston Nutter had commented upon the presence of five different heads. He had specifically mentioned the head of Frank Miller as being among the five at the "massacre site." He also said that four of the skulls were marked by a hatchet, and one had been caved in from a fierce and heavy blow. Consequently, I wasn't prepared to give a definitive statement about the number of skulls we would find until I exhumed the remains of the prospectors. I didn't realize then that the skull articulates so loosely with the atlas (the first cervical vertebrae) that when there is no flesh to hold it in place it is subject to being separated from the rest of the body (the postcranial skeleton) by the slightest impact from a marauding carnivore or other, even moderate, force. As a result, it is not uncommon to find a defleshed skull separate by a considerable distance from the body to which it was attached in life.

This was my first exhumation and I was learning fast, but my education would most assuredly improve on the marrow.

SCIENCE HAD NEVER BEEN UTILIZED FOR ANY PURPOSE IN THE two trials of Alfred Packer. No bullet or firearms identifications were conducted, nor would such identifications have been expected at that time and place. No assessment of the trauma inflicted by the hatchet blows or by any other sharp-force instrument took place. The bones of the five deceased prospectors were not examined with an eye to the details of interest to a forensic anthropologist. Not a lick of scientific analysis appears in the transcript of the first trial or in what we know from the reports of the second trial, for which a transcript is among the missing. Indeed, it was not even known for certain whether Packer's admitted cannibalism demonstrated that he, in the way of the Eskimos, ate the flesh of his dead companions raw or if he, unlike the Eskimos, cooked the flesh either on or off the bones from which it was removed. To put it harshly, he was either a ribs man or he was not.

It was a documented fact, however, that he was in possession of a hunting knife when he emerged at Saguache in the spring of 1874. That knife became the prime focus of my aboveground investigative attention.

The missing knife was likely to answer one unresolved issue when the bones and the knife were both examined—namely, were the victims' bones defleshed and, if so, was the missing knife, possessed by Packer at Saguache, the culprit in the defleshing? Furthermore, the cut marks from a knife found on the bones were, at least in theory, subject to being associated with a particular knife through unique striations left by the knife blade.

The media attention given to my project caused many knives to be submitted to me as candidates for the Packer knife. It was at a lecture by me, sponsored by the GWU law alumni association at Baby Doe's restaurant in Denver, that the Packer knife unexpectedly surfaced. Its owner courteously left it in my custody for closer analysis in the laboratory at my university. He also provided me with his recollections of the knife's provenance that would account for its history of ownership and his possession of it.

As he explained the knife's history, it had come into his hands from the Colorado Springs sheriff's office. The sheriff there had been charged with planning the execution of Packer as ordered by Judge Gerry at Packer's Lake City trial in 1883. That same sheriff came into possession of the knife that I had been given to authenticate. Years later, when the Colorado Springs police moved to new quarters, the knife was transferred to a crime reporter for a Colorado Springs newspaper. And that reporter passed it along to his son, who now presented it to me for analysis.

Part of the tale of the knife's ancestry included its having been kept in a gun-display case in the home of the present owner at a time when the house was burglarized. The burglars took all the guns that were in the display case but left the knife, sheathed in a leather scabbard, behind. Little did the thieves realize the value of the item they had overlooked, once connected to Alfred Packer and his cannibalizing the remains.

My laboratory analysis of the knife, including dismantling it to look for concealed traces of blood, was unavailing in my authentication efforts. But Walt Birkby, the team's forensic anthropologist, came to my rescue, as he so often did in the course of this project. He was surely the rock upon which the project stood.

Walt, in his inimitable cut-and-dried way, turned the knife on its side and in a moment's examination pointed out some scratches on its steel hilt. Under the microscope those scratches turned out to be the three letters "WFM" in block printing. At least two of those letters matched the initials of Frank Miller, one of the five prospector murder victims. But without further data, Frank Miller's ownership of the knife could not be even probabilistically affirmed.

That additional data was shortly forthcoming. In the files of the National Archives in Washington, D.C., all Millers listed in the 1870 national census who were then living in Salt Lake City, Utah, were excerpted by me. The only Miller even remotely connected to the Packer party of six prospectors was a Miller who had emigrated from Germany. Tradition had it that Frank "Reddy" Miller had been a native of Germany who had come to the United States. This constituted some added circumstantial proof associating the knife with Miller, but it did not suffice in my mind as decisive.

Checking further in the 1870 census records, I found that a number of the original twenty-one prospectors who had left Utah with Alfred Packer in 1873 were also listed as having resided in the same lodgings in Salt Lake City in 1870.

And then, mirabile dictu, they were found to be housed at the same place as recorded for Frank Miller, whose place of birth was listed in the census as what is present-day Germany. But my glee over being now able to prove the knife's ownership by Frank Miller was cut short by my astute research assistant, Sherrie Hardwick.

Once again I had missed the obvious. Walt Birkby's observant eye seemed to be shadowing me, now in the person of Sherrie Hardwick. "But Professor," she said deferentially, "we know it was Frank Miller who died at Dead Man's Gulch in 1874."

"Yes, yes," I replied impatiently.

"Well," she continued, "the census report registers the presence in 1870 in Salt Lake City of a Wilhelm Miller, not a Frank Miller."

Momentarily taken aback by her pointing to my being too quick to find Frank Miller in the census records, I took some seconds to fashion an unreproving reply: "Yes, but Miller's status as an immigrant from Germany tells us more than that. It also suggests that he changed his name from the Wilhelm Miller appearing in the census to Frank Miller, the man buried in the Jarmans' front yard. It was most probable that the 'WFM' scratched into the knife's hilt were the initials for Wilhelm Frank Miller, whose trade was known to be butcher. After all, is it unlikely that a butcher would possess a knife with his initials etched into it?"

The confluence of this new data sufficed for me to report that this knife bulked largest as the most likely candidate for the one Packer used to cannibalize his five prospector companions. But further scientific investigations following the exhumation would provide additional assurances.

The questions that we wanted the exhumation to resolve included whether there was clear evidence of violence against these men as opposed to natural death (starvation and/or hypothermia), and whether we could determine if they'd been defleshed in a cannibalistic endeavor. If the manner of death was homicide, what was the instrument that caused the deaths, and was Alfred Packer to be targeted as the sole killer or only one of the killers?

We would assess the pathology of the bones and teeth, look for blunt-force injuries or bullet wounds, search for metallic pathways through the bones, look for knife or hatchet marks, and use chemical analysis to examine trace evidence of potential gunshot residue, the imputrescible and then ubiquitous black powder being the ammunition source. We alerted the Hinsdale County Museum that any artifacts discovered would be placed in their care.

It was possible, I knew, that the many naysayers would be right and we would find nothing at all. It had been 115 years since these men had been buried, and no one was even certain of the exact location of the grave. We had a memorial marker, to be sure, but there was also some dispute about whether the marker was even in the correct place. I thought about how I could minimize the amount of digging we might have to do. I didn't want to think about days, even weeks, of effort with nothing to show for it. Yet we were already in motion. There was no turning back. As the dig approached, I became ever more nervous.

Then I read about a technology that might be of enormous assistance in quieting my nerves.

As I read through scientific magazines in the disciplines of anthropology and archaeology for items deserving of mention in *Scientific Sleuthing Review,* I had spotted a magazine advertisement for a remote-sensing technique known as ground-penetrating radar (also known as subsurface interface radar). The company that manufactured and marketed it was given as Geophysical Survey Systems, Inc., in Hudson, New Hampshire.

The ad mentioned a variety of different projects to which its use was suited, such as finding burst pipes beneath city streets. Nothing was said about its utility in locating buried human remains, which caused me to wonder whether it could be helpful to the Packer project. I did more research, and what I read about the process struck a responsive chord. I called the company in New Hampshire and spoke to Stan Smith, one of their field engineers. His equipment had never been used to locate buried bodies, but he agreed that it would be a wonderful opportunity for them to market their equipment in a different venue. To my delight, they were willing to bring it out to Colorado and—will wonders never cease?—at their own expense.

All was now in readiness. We had lodgings at the spacious and luxurious Lakeside Lodge, which sadly no longer exists, overlooking Lake San Cristobel. The Murphys, as proprietors, would generously catch fresh trout in the lake each morning for our breakfast. In the evenings when we returned from our labors in the sun, cocktails would await us. That hospitality uplifted our spirits, although as it turned out we didn't need much help of that spiritous sort. The spirits of a different but equally intoxicating kind were with us from the outset.

We began our exhumation on Monday, July 17, 1989, quite fortuitously the hundredth anniversary of the birth of the creator of Perry Mason, that lawyer sleuth who knew no equal, as well as quite deliberately the fourth birthday of my grandson, Willie. Before a backhoe or a shovel broke the ground, Stan Smith put his subsurface interface radar system (SIR System-8) into operation. Prior to that, our engineer Jim Packer had readied the area with a grid map, and Byrne Smith had performed his backhoe magic in removing the boulder and the fence posts surrounding it.

The ground-penetrating radar (GPR) system relies on high-frequency electromagnetic pulses to plot an echoed profile of the subsurface structure, showing anomalies in a distinctive parabolic shape. In other words, it measures the depth and shape of things (termed "anomalies") under the surface, whether buried or natural to the subsurface. If we could determine exactly where the remains lay, we could zero in on the grave-site area and bring the backhoe into operation.

Jim Ayres, the team's archaeologist, a seasoned professional, thought this step was absurd. He was reluctant to attend the site while the GPR operation was under way. But Walt Birkby, our anthropologist, was open-minded on the subject and interested in seeing what this new technique could accomplish. He and his doctoral students, Todd Fenton and Bruce Anderson, who have gone on to careers as forensic anthropologists, were present at the site while the GPR operation was being undertaken.

The traditional method of proving the presence of a grave is to remove a few inches of topsoil, whether by shoveling or through a backhoe. Then any discoloration in the area of disturbed soil for a grave site can be compared to the original color of the undisturbed adjacent soil. When digging a grave, various zonation levels of soil are penetrated and mixed upon refilling the pit after the burial.

The topsoil is typically organic and very dark, but as you go down, the color changes from dark to light. Then, when you put all that soil back to fill in the hole, you've mixed the different levels, so the coloration is distinctively different from that at any other undisturbed place and will remain that way forever. With airborne infrared photography, the southwestern trails and burial sites of the Anasazi Indians stand out in bold relief.

The mixed coloration of disturbed soil is only one of the archaeological tools that can be used. Another requires scanning the soil at ground level with an eye for soil subsidence. When a coffin collapses or when disturbed soil settles, a declivity in the ground appears that in established cemeteries is filled with a sprinkling of topsoil. There was no soil subsidence to be observed on the grave site at the Jarmans' property. Therefore, we were entirely reliant in the first instance on Stan Smith's GPR and secondarily on Byrne Smith's backhoe with a straight, never a toothed, bucket. (The teeth have a ripping effect that can do terrible damage to a coffin and/or buried human remains.)

As I watched Todd Fenton walking the transducer over the gridded area while Stan Smith watched at the monitor screen and noted the printout that was being produced, I indulged myself in enjoying this crisp summer morning in the mountains with blue skies and a good weather prognosis for the entire week. Breathing in the fresh air, I watched and waited.

Stan had set up his radar-control circuitry, graphic recording capabilities, and two high-frequency transmitter/receiver antennae. Todd ran the antennae over the mapped grid pattern in the area where the boulder-memorial stone had been allowing the graphic recorders to document the results.

Strong signals came out black, weak signals were white, and intermediate signals recorded as gray. Stan watched the graphs and monitor intently, and when he brought the antennae around the monument's former location he announced, "There is an anomaly very close to the surface, maybe twelve inches under."

"Twelve inches?" I asked. I knew it was possible, but it seemed too quick and easy to be scientific. Yet good science, I knew, is not always the product of the sweat of the brow or the aching lumbar spine from bending over a microscope.

"I don't think you should use the backhoe here," he warned me. Whatever the anomaly he had detected might be, he strongly recommended not chancing the damaging effect of the backhoe. As a result, Byrne Smith was put off for another day. I announced that shovels were now the order of the day in exposing the area of the exhumation.

It took no more than a few short hours for one of the team members to sound the cry that a bone had been found. Upon inspection, Walt declared it to be a fragment from a human parietal bone.

I was overjoyed and gave thought to having a victory party that night at the lodge. But Walt dimmed my enthusiasm by noting that the parietal bone is one of the most sturdy bones in the body, meaning that its having been found did not necessarily mean that other, less robust bones would be in similar undecayed condition.

However, Walt's prudent pessimism was for naught. Shortly, more bones were found until it became apparent that the skeletal remains of the five prospectors were being paintbrushed by us into public view. Happily, the bones of the five were not commingled as they would have been in an ossuary. As each of the five prospectors was exposed, the bones revealed five separate and distinct burials, all cheek by jowl with one another. None of the prospectors seemed to have been buried in a coffin or other container. Fibers akin to burlap adhered to some of the bones, indicating that the bodies had likely been bundled in burlap and carried up the incline from the Lake Fork of the Gunnison River and deposited in a shallow grave on the level ground above the river.

Toward the end of our first day of digging, with bones being uncovered in large lot numbers, the visual media, with their satellite dishes at the ready, converged on the site. The word had somehow been circulated that the dig was a resounding success, so far at least as to the bones being found in fine fettle for our closer laboratory work.

I was approached by a television newsman from a Denver station who wanted a live interview for the 6:00 P.M. news. He was interested in a statement concerning the bones we were unearthing. I sought to redirect him to Walt Birkby, our team's specialist in human bones, who was standing off to the side of the excavation pit.

The newsman looked briefly at Walt, seeing a slim, trim, and wiry ex-marine still proud of his short-cropped marine haircut. Then he turned to me and gave me a similar once-over, noting the floppy hat I was wearing and my short-clipped gray beard.

"No," he declared. "I'd rather talk to you. You look like the father of Indiana Jones." Realizing that Sean Connery had recently been seen in that role on a new Indiana Jones flick, I succumbed to the temptation he offered me.

So much for the media's interest in presenting the most informed perspective on a subject, particularly a scientific one. This encounter has been repeated many times over in my subsequent appearances on television—with or without a lab coat, as the media may designate. In respect to catering to image, the print media are light-years ahead of television productions, because the reliance is not on visuals but on the language arts.

It now was time for a rigorous protocol for documentation and recovery to be in place to ensure the scientific acceptability of our findings. For our project manual, I had included an article that Walt had published in 1978 in the *FBI Law Enforcement Bulletin* about the process of forensic exhumations. In it, he and coauthor Dr. William Bass provided important guidelines. They urged that photographs be taken of each stage of the project and that identification markers be used. Where more than one individual was found to be placed in a single grave, as we had here, each individual was to be given a separate letter designation. Walt therefore identified the prospectors as A, B, C, D, and E, according to the order in which they were exposed. Our careful and conscientious work of paintbrushing the soil from the bones continued for the rest of the day.

That first evening, we returned to the hotel for our celebratory cocktail party. My elation knew no bounds. I looked out at Lake San Cristobel from my room at the lodge and exclaimed to the unlistening world, "I did it! I did it." In my enthusiasm I jumped up, pumping my fist in the air, and it went right through the ceiling of the room. The other team members shared my elation, although with less damage to the property of the lodge.

The next day and those following, the work became painstakingly intense. To our relief, as we carefully brushed the soil from the skeletal remains, we saw that the victims had actually been placed side by side. That positioning eased the burden of our task, as did the fact that the bones turned out to be in excellent, well-preserved condition. They were left, as per the rules of the archaeological process, in place until we were able to expose all five over the course of the next six days.

The media attention to our work was overwhelming, with satellite dishes and the like crowding the roadway to the burial site. Children were drawn like magnets to the edges of the pit where they could see the bones of the prospectors being uncovered. The parents, a little less enthusiastic, hovered in the background. On one occasion, a *Los Angeles Times* reporter was said to have entered the pit to have a closer look at the bones. I confronted her and instructed the sheriff to escort her from the exhumation site. Needless to say, she had her revenge, the power of the press being what it is. The next day, the *Los Angeles Times* printed a story by her, mocking me for strutting about like the father of Indiana Jones and playing Sean Connery to the crowd. I had no recourse but to take due note that offending the press can create a no-win situation for the victim of the press's outrage.

As careful as I try to be, especially with members of the media present, I can't control all the members of my team. One unfortunate incident involving a scientific member of my team and the press haunts and plagues me even to this day. On the second day, as we were in the process of uncovering the remains, a forensic pathologist with our team thought he saw something unusual on a hipbone (an innominate) from one of the prospectors. He picked up the bone, drew attention to a hole in it, and precipitately and unguardedly blurted out, "A bullet hole! A bullet hole!"

To my utter distress, the next day's headlines in the *Denver Post* read, "Alfred Packer Exonerated. Bullet Hole Found." We hadn't even begun our analysis, and here we had reporters closing the case due to a slip of the tongue of a team member. Slips of the tongue do more than sink ships. They also jeopardize the credibility of scientific investigations conducted in full view of the public and the media.

After a more careful examination, the hole in the bone was found to be not a bullet hole at all but a carnivore's (possibly a wolf's) gnaw marks. Apparently, the defleshing of the bodies was not only the business of Alfred Packer. The bone (as was the case for at least three other hipbones from the other victims that we subsequently detected in our Tucson laboratory analyses) bore the distinctive and unmistakable signs of the teeth marks of an animal and none of the beveling that you'd expect to find on the exit side of a bullet hole. It was unquestionably not a bullet hole, but the damage had been done and it could not be undone. I had learned another salutary lesson in orchestrating an exhumation: The only member of the team who should describe the progress of the project to the press is the project director himself.

Another vital lesson was also in the making. As the bones were exposed from their soil-protected closeting, the sound of cracking and popping was heard. The heat of the fiery sun was drying the bones and causing them to crack, creating the possibility of artifactual bone fractures that could be mistaken as having occurred at the time of death. Quickly, the bones were covered with a tarpaulin and the word went out for a tent to be placed over the site.

Another unexpected incident occurred once and not a second time. On the morning of the third day of our digging, when we returned to the grave site we found a mess of small bones strewn about. It didn't take an anthropologist to see that they were chicken bones that some joker had deposited at the site during the previous night. I decided upon a course of action to stem the tide of this pranksterism.

I asked our team's engineer, Jim Packer, to stand by me as I addressed the assembled crowd at the grave site. I remonstrated about the need to preserve the integrity and the sanctity of the grave site against childish marauding. In order to protect the site, I declared that Jim, all six feet and more of him, along with his large-boned Irish wolfhound, would be tenting out at the site. In addition, I intoned that if there were any further chicken bone incidents, Packer—pointing to the impressive figure of Jim

Packer—would get you. I do believe the crowd got the message, since I saw no smiling faces in response to my threatening words.

Once the remains were fully exposed, Walt sketched them in their respective positions in the grave, much like what had been done in 1874 by John Randolph when they lay aboveground. This way we had something to which to refer as we reconstructed the events surrounding the burial. This place was for all intents and purposes a crime scene, which we wanted to treat in a fully scientific way so that our work product would gain acceptance with other scientists.

The removal of the bones was laborious and necessarily exacting. Five boxes were prepared, labeled A, B, C, D, and E. The bones for each designated person were placed separately in each box as Walt marked off on anatomical charts the names of the bones that were being boxed. There was one glitch, however.

A tree root had sent its leaders into the skull of Shannon Wilson Bell, giving us trouble in separating the skull from the tree root. The root was cut with a pocketknife, but the knife slipped and left a fresh cut mark in the skull. We now had an additional cut mark to explain, somewhat shamefacedly. Even with the best of care, such things happen and must, in spite of the embarrassment, be documented.

There we were with the bones of these deceased prospectors from over a century before neatly boxed and ready to travel in my Chevy Blazer to Walt Birkby's laboratory in Tucson, Arizona. With the new skills and methods of twentieth-century scientists, we were about to turn our attention to solving the mysteries of an age-old crime. Even a casual and cursory inspection of the bones in the pit and as we boxed them told us that our closer laboratory analysis would be rewarding.

It seemed pellucidly clear, even at this early phase of our investigations, that Alfred Packer was no hero, nor was he innocent of all wrongdoing, as some had said. Someone had defleshed these bones, and that someone was by all lights Alfred Packer. If so, he could be counted as a liar, a murderer, and a cannibal. Still, our official interpretation would have to wait for our more fulsome and penetrating laboratory analysis of the bones. We bid a temporary adieu to our wonderful mountain lodgings, our cocktail parties, and our trout breakfasts, and caravanned off to Tucson, Arizona, where long and tedious hours of work awaited us all.

The trip to Tucson was uneventful, our caravan deciding that the condition of the prospectors' remains should not be put to the test of a shorter but bumpier ride over the mountains via Durango, Colorado, to Tucson. There was one indecisive moment when, at our overnight motel, the matter of where the five boxes of bones would sleep was raised, either in one of our cars or inside one of our rooms. It was uncomplainingly agreed by all that the prospectors deserved the consideration of resting in the company of their protectors. We slept in peace in that knowledge. And they, the prospectors, for their part slept in pieces.

The next three weeks were engaged in cleaning, piecing together, and examining the bones, including taking selective X-rays. The so-called bullet hole in the hipbone of Shannon Wilson Bell was definitively ruled to be a carnivore's claw mark. This hole might have been the same one that Preston Nutter had seen and called a bullet hole without giving its exact anatomical location.

The bones from all five sets of remains demonstrated clear and decisive indicators of violent deaths and of being substantially defleshed, with only the skulls and the pubic regions having escaped the work of the defleshing knife. Shannon Wilson Bell was probably victim C, the man splayed out in the middle of John Randolph's *Harper's Weekly* sketch and also the man positioned in the middle of the grave pit we had unearthed. Bell, as depicted by Randolph's sketch as well as our victim C, had many features lacking in the other four victims. The cantilevered lower leg, the outstretched right arm, and the tilt of the head were identical in both Randolph's depiction and in victim C. He was in all respects the man in the middle in this controversial slaughtering.

The four skulls we had found bore the marks of repeated blunt-force trauma, some of which were evidently made by the blade of a hatchetlike implement. It had been a barrage of blows, not a spontaneous act, as Packer had described, to finish off Bell.

Defensive wounds were the order of the day on all the bones, save those of Israel Swan. The cut marks on the arms and hands of the bones told the sad tale of the prospectors' last moments as they sought to ward off the death-dealing hatchet blows of their attacker. They all seemed to be right-handers, from the wear marks in the shoulder girdles of each, and, as expected, the defensive wounds were all sustained ipsilaterally, in their off hands and arms. The wounds we found were those typically seen in such circumstances, where the less dominant limb is sacrificed in the act, futile as it was in this instance, of seeking to ward off the attack, saving the dominant limb for self-defensive maneuvers.

We took still photographs of the bones and the perimortem (time of death) and postmortem (after death) cut marks on the bones to document our findings for posterity. Every available scientific course of action was pursued save DNA testing. At that time, DNA profiling (then misleadingly termed DNA "fingerprinting") was just coming on board for forensic uses. Also, we lacked any relatives of the five prospectors to provide the necessary reference samples (exemplars) for comparison purposes.

Toward the end of our three-week analyses I called upon Luke Haag, a prominent tool-mark examiner from the Phoenix area, to examine the cut marks on the bones so as to determine whether one knife as well as one hatchet had been used to inflict the various marks on the bones that we had detected. As always, Luke gave his unstinting, thorough, and immediate attention to my request, including Mikrosil casting (a Silly Putty type of material resulting from mixing a fluid compound with a mass of grayish, soaplike material). The casting effectively reproduced the fine striae embedded in the cut marks so that they could be examined microscopically.

Unfortunately, the results of Luke's examinations did not contribute any dynamite findings. He could not exclude the possibility that either multiple hatchets or multiples knives had been employed either in the killing or the defleshing. The limitations imposed upon his scientific work did not arise from the poor condition of the cut marks after so many years in the ground. Rather, he was curtailed by the lack of a known reference sample (knife or hatchet) to compare to his Mikrosil casts. In addition, the knife marks were more often than not of the filleting type and not the gouging type most suitable for the association to a known knife.

Indeed, Todd Fenton, Walt Birkby's graduate student, was so careful and engaged so much time and effort in documenting and counting the 478 individual cut marks on the bones of the five victims that we began terming him our "cut mark man." With that labeling you can imagine the puns that resounded day after day in Walt's lab: "Todd, you are a 'cut' apart." Todd, then our only unmarried team member, was peppered with comments about his after-hours doings, such as "Don't be a cutup tonight, Todd," or culinary admonitions such as "Try a fillet at a restaurant tonight, Todd." And, of course, none of us was "cut" from the same imperturbable cloth as Todd Fenton.

After the analyses were concluded, our reconstruction of the death of the prospectors singled out Packer as the predator who had treated his five companions as his prey. We conjectured that he had come upon some or all of them by stealth, as they slept the fitful sleep of starvation and hyperthermia, and hacked their skulls with murderous intensity.

Israel Swan, a fiftyish Canadian emigrant, probably put up little or no fight. There were no defensive wounds on his arms or hands, and his skull was battered by the blunt force of a hatchet blade in only a few locations sufficient to cause the trauma from which he died. Pathologists denominate such wounds to the head as "cephalic insults."

Records at the National Archives revealed that Swan had served in the Mexican-American War with the famed Donephan's Brigade. He joined Donephan's volunteers in St. Louis, Missouri, and became a part of what, in military circles, was an epic journey, fully the equal of General George Patton's race into Germany.

Donephan's troops, like Patton's, lived off the land they traversed, but unlike Patton, Donephan never lost a man in battle. The records of Donephan's expedition include a reference to Israel Swan's having been taken temporarily ill on one occasion.

Whereas Donephan went on to become governor of Missouri, Swan cashed in his land-grant award for cold cash to enable him to play a role in the California gold rush. Whatever possessed Swan, a seasoned veteran trekker who with Donephan had crossed the snow-laden mountains of southern Colorado and who would have been severely incapacitated by his advanced spinal arthritis, to have thrown in his lot with Alfred Packer is a mystery well beyond answering with the facts at hand.

While Israel Swan, in his weakened physical condition, may have been dispatched in his sleep, the other four prospectors were probably not.

Every one of them bore the talismanic signs of having defended himself against a hatchet-wielding attacker. People are known to have put a hand to the muzzle of a gun to stymie an attack when, upon calm reflection, a hand would be known to be unable to stop a bullet. But at the dire moment of a fatal attack, calm reflection is supplanted by a wild hope and an instinctive response that sacrificing a limb will stave off an otherwise inevitable death.

James Humphrey, Shannon Wilson Bell, George "California" Noon, and Frank "Reddy" Miller knew that they were defenseless and that they were under attack by a murderous hatchet-wielder. Israel Swan was out of the picture, weakened by age and infirmity, and quickly killed. The others, however, defended themselves with varying degrees of vigor and determination.

How do we know this as a scientific fact? The marks on the bones spoke tellingly of defensive wounds. Frank Miller and James Humphrey were bludgeoned and butted into eternity by an ax that was not dissuaded or obstructed by the arms and hands Miller and Humphrey raised in futile self-defense. Their skeletonized limbs told us that much. The defensive wounds were many and patent, and they were all caused by an axlike weapon.

George "California" Noon put up quite a struggle to save his life. Only eighteen or so years of age (evident from the partial epiphysis—closure—of the bones of his left humerus) and an Irishman (my Irish heritage coming to the fore), he would likely not look kindly on the prospect of his death at the hands of a hatchet-wielding aggressor. He did not go to his death without a fierce struggle, as manifested by the complete severing of his left humerus, leaving it in three separate pieces that we recovered, and the multiple wounds to his skull, all inflicted by a hatchetlike implement. The stub of the severed humerus can be seen, still attached to Noon's body, in Randolph's sketch. His bones spoke with a voice of anguish about his final earthly moments.

It was not possible for us to tell the order in which the five prospectors were hacked to their death. However, Packer maintained in one or more of his many statements, and at his Lake City trial, that he had killed Shannon Wilson Bell when he returned to their Lake Fork riverside encampment to find Bell cooking the flesh of Miller, the butcher.

Bell, according to Packer, attacked him with a hatchet, not a rifle, which Packer claimed to have in his hands at the time. The only gun said to have been possessed by the five has been claimed to be a Winchester rifle belonging to Noon. The rifle has never been brought to contemporary light.

Packer claimed that "when the man [Bell] saw me, he got up with his hatchet towards me when I shot him sideways through the belly, he fell on his face, the hatchet fell forward. I grabbed it and hit him in the top of the head."

Packer's recital can most fairly be described as a cock-and-bull story. Why did he feel it necessary to kill Bell with the hatchet when Bell had been immobilized by the bullet he shot into him? Why did he not shoot another bullet into Bell, assuming he still needed to act in self-defense?

Moreover, our examination of Bell's bones told a wholly different tale, with Bell acting in self-defense against a hatcheting aggressor. How do we know that? The defensive wounds on Bell were determinative. His left humerus suffered such a tremendous hatchet blow that it was almost cut through. His left hand had hatchet marks on the back side of the left third and fourth metacarpals (hand bones), the result of a separate blow or blows from the hatchet.

Further, Bell was not merely struck by the hatchet "in the top of the head," as Packer professed. Bell's skull showed the indisputable marks of multiple hatchet blows into and about his face.

In point of fact, Todd Fenton called my attention to five separate sharp-pointed-instrument (probably hatchet) blows to Bell's skull as well as two distinct sites of blunt-force trauma on the left side of his skull. It was possible to reconstruct Bell's killing by Packer to demonstrate Packer's striking Bell's skull first with the sharp edge of the hatchet from one side and then, in a return blow, like the backhand of a tennis player, battering him with the butt end of the hatchet from the other side of his victim's skull.

Little did Packer realize that the bones of Bell would in 1989 put the lie to Packer's story of Bell's death, causing Bell's voice to cry out the truth from the grave.

The laboratory efforts now being at an end, a five-compartment box was crafted as the reburial container for the skeletal remains. It is always a source of some mystification and wonderment for me in these projects where skeletal remains are involved how little space the human skeleton occupies when packed into a box. The trip back to Lake City from Tucson was a one-day trip, since I traveled off-road over the high mountains, scaring the bejeebers out of me but not stoical Bruce Anderson in the process. The Lakeside Lodge was ready for our return.

Before the reburial, I undertook to perform an experiment of my own. I wanted to retrieve bone samples from each victim for Doug Ubelaker at the Smithsonian. He was using a new type of bone aging known as osteon counting. Osteons are small circular bone structures that form throughout the developmental stages of life. In human adults, they're randomly scattered throughout the bone's cortex, which distinguishes the bone from that of an animal, in which osteons line up in rows. The idea of counting them to determine the age of a skeleton was developed in 1965 by anthropologist Ellis Kerley, then revised with Ubelaker's assistance. It was based on our knowledge of the human growth and development process. Supposedly, there are more osteons present with advanced age, and so to calculate age they counted the osteons in questioned bone samples.

Ubelaker and Kerley had authored articles in scientific journals propounding that osteon counting held promise as a scientific technique for the aging of bone. With Doug's concurrence, I decided to conduct an experiment on the relative merits of osteon counting in contrast to the traditional bone-aging technique, which is dependent on degenerative osteoarthritic changes, most particularly in the costal cartilages of the proximal (near) ends of the ribs, that occur naturally with aging.

Walt Birkby was a believer in the costal cartridge method, and Doug Ubelaker agreed to try the osteon route. After the Birkby and Ubelaker results were reported to me, it would then be time to evaluate the respective outcomes. It was a limited but intriguing opportunity that I thought to be timely in a scientific sense.

But the experiment hinged on my obtaining bone-section specimens from each of the prospectors' long bones. The only catch to that was that Walt was so committed to the costal cartilage aging method that, to him, osteon counting was totally valueless.

To circumvent Walt's probable objections, and proceeding without his knowledge, I bought a ten-inch brass Sheffield steel straight-back saw from a Lake City hardware store (the usual electrically operated autopsy saw, generally a Stryker saw, not being available to us) to use in securing femoral sections from each of the prospectors while they rested boxed and ready for reburial in my room at the Lakeside Lodge.

With the assistance of Bruce Anderson and Todd Fenton, stationed as lookouts in the hotel's long hallway leading to my room, I sawed away unimpeded by anything more than the fear that I would be interrupted by Walt's arrival. Fortunately, my luck held and the bone sections, marked A through E, were secured. Following the reburial, they were delivered by me personally to Doug's laboratory at the Smithsonian Institution in Washington, D.C.

The femoral sections having been obtained, the bones from which they were taken were placed back in the five-compartment coffin. The lid was closed and all was now in readiness for the reburial ceremony, which was interdenominational, with clergy from each of the religious faiths resident in Lake City officiating. Each minister either read from the Bible or delivered a reburial homily suited to the occasion. Not being in possession of any sure knowledge of the faith of any of the five persons we were reburying, all agreed that such a solemn ritual was preferable to none at all.

The solemnity of the moment was broken, indeed smashed to smithereens, when it came time to close the burial pit with backfilled soil. The soil had to wait, for the Jarmans had other quite eye-opening and unannounced plans for permanently sealing the grave from any future prying eyes or grave diggers. The situation was reminiscent of the burial of American gangster and Public Enemy Number One John Dillinger on July 25, 1934, at the Crown Hill Cemetery in Indianapolis, Indiana.

Today it is said a large headstone bearing John Dillinger's name stands on his burial plot, but only a few are aware that Dillinger's father, also a John, decided, in his grief over the killing of his favored son, to ensure that no one would ever tamper with Dillinger's grave or, more important, his mortal remains. In that frame of mind, Old John did what the Jarmans did at the grave site of Alfred Packer's fellow prospectors.

Whereas Old John poured yards and yards of concrete over the coffin of his son, the Jarmans had Byrne Smith, the team's forensic backhoe operator, lift a large and heavy steel plate, almost precisely fitting the size of the reburial grave, into the air over the open pit containing the five-compartment, specially constructed coffin and then drop it noisily and unceremoniously into the pit. The sound of the plate's hitting the coffin was more heartrending than deafening. After weeks of toil by us in Walt's Tucson laboratory cleaning, analyzing, and applying a lacquer preservative to these bones, all those exhausting efforts were in that one moment of the steel plate's descent wiped out.

We could not hear the fragmenting and crunching of the bones in their perishable wood coffin, but we felt it in our bones deeply and profoundly. We had come to know and to share in the anguish of these five prospectors, which anguish we were called to relive at this moment. Human emotions do not evaporate in the presence of human remains, skeletal or otherwise.

Weeks later, after dropping the bone section off with Doug, I called him to check on the progress of his osteon counting. I asked him, "How is the osteon work coming along?"

"All right," he answered. "What did Walt find with the costal cartilage?"

"No, no, no," I said. "I can't tell you that. This is a blind trial of your osteon method. You tell me."

"Well, I found that victim E [Noon] is about thirty to thirty-five years old, give or take some years."

I told him that from the almost fused growth plate on Noon's left humerus he was about eighteen years old, and that as to him we didn't need to look to the costal cartilages. At that point it seemed, at least as to Noon, that the osteon counting was wide of the mark and not scientifically reliable. Yet when I received Doug's official report on the samples, it referenced sample E (Noon) as an eighteen-year-old. Doug explained the discrepancy by saying he had recalculated his original evaluation.

The end of the trail of Alfred Packer was not yet in sight. The tales linger and fester and burst forth with new vigor, as they did after my results were announced at the National Press Club in Washington, D.C.

A Packer apologist, said to be a museum curator from the western reaches of Colorado, came forward with the entirely fanciful tale that his new scientific discoveries proved that Packer acted in self-defense, as he had said, and that therefore his innocence had been proved.

This new claim relied on the supposed finding of a handgun at or near the grave site in Lake City as well as gunshot residues discovered in the soil, the two being connected to each other and to the killings, so it is claimed. On their face, these assertions were preposterous.

We had screened the soil as we removed it with a top-of-the-line Garrett metal detector. We found bottle caps, Civil War buttons, and metallic bric-a-brac, but no gun or other metallic objects that could be associated with the killings. In addition, our examination of the bones did not give any evidence at all of gunshot-related trauma to the bones, except for a pseudo-arthritic joint on Bell's left arm.

The radiographs of Bell's left arm showed evidence of traces of lead from the passage of a lead bullet. However, those radiographs also demonstrated that that gunshot injury had been sustained many years before his death. The remodeling of the untreated injury had left Bell with restricted motion in his left arm, an arm lacking in the radiopacity sufficient to show up as bone under X-raying. The body had mended the wound with false cartilage, creating a pseudo-arthritic joint wholly unrelated to Bell's death.

Additionally, the considerable literature on the Lake City deaths made no mention of the presence of a handgun, nor does any such weapon appear in the testimony, such as exists, from Packer's two trials. The accounts describe a rifle, probably the Winchester rifle owned by George Noon that Packer carried off with him. Maybe that handgun is the Harrington & Richardson handgun involved in the Sacco and Vanzetti robbery-murder in South Braintree, Massachusetts, which has been long unaccounted for. If truth is to be stretched to the breaking point, why not stretch it even further?

With my research at an end, I could rest comfortable in the knowledge that I did manage to correct at least one long-standing error of Packermania. The student dining hall at the University of Colorado at Boulder was for years called the Alfred E. Packer cafeteria, perhaps because of the quality of the food dished out there. Yet they had his middle initial wrong. After I reported our results, they changed the name of the cafeteria to the Alfred G. Packer. Of all the laborious work I did in this investigation, this name change may be the only permanent and unassailable rectification of the historical record. Such are the risks of taking on the many legends of a most undeserving mythic man.

In my opinion, our new and incontrovertible scientific facts supplement the case made by the prosecution against Packer in both of his trials. I am convinced that our evidence convicts him of homicide not only beyond a reasonable doubt, as two juries decided, but also beyond any scintilla of a doubt.

But I fail in my task of reportage if I do not give notice of the terra incognita of this case. Every belowground investigation, whether exhumation or excavation, in which I have been involved has a surprising, unlikely, and unknown ingredient to cap the story of our work. That of Alfred Packer is no exception.

In tabulating the cut marks to the bones of the prospectors accountable to Packer's defleshing them, it was not only noted, as previously mentioned, that the skulls and the pubic regions were exempt from traces of cut marks. It was also discovered by Todd Fenton, in his counting and locating the cannibalistic cut marks, that the overwhelming majority of the marks were to the posterior of the victims, not to the front of their bodies.

My immediate reaction to this statistical information was to remark jocosely that apparently Alfred Packer did not believe in the well-worn canard that the breast meat is the sweetest meat known to man. On more serious reflection, I decided it was more probable that Packer got a heavy, restraining twinge of a guilty conscience when he removed flesh from his victims while they were lying supine looking him in the eye. So he was a cannibal and a killer, but he had a conscience after all—sort of.

Now that I had completed this project, my appetite, after the fashion of Packer, was whetted for another such engrossing adventure. I went in search of another historical controversy in need of scientific attention by way of an exhumation. The wait was not prolonged.

2. CARL AUSTIN WEISS, M.D.

He Died in Marble Halls

O they say he was a crook
  But he gave us free school book
Now tell us why it is that they kill Huey Long?
  Now he's dead and in his grave
But we riding on his pave
  Tell me why is it they kill Huey Long?

CAJUN BALLAD

The assassination of a political figure is most often executed by persons who seemingly have nothing to lose by their actions. Lee Harvey Oswald, Sirhan Sirhan, and James Earl Ray, the assassins of President John F. Kennedy, Senator Robert Kennedy, and Dr. Martin Luther King, Jr., respectively, fit neatly into this category of down-and-outers. But Dr. Carl Austin Weiss, reputedly the assassin of Louisiana senator Huey P. Long, was not of that mold.

Indeed, he is a rarity among proved or purported assassins—a medical doctor. He is the only physician in history to be an accused assassin. To add further mystery to this case, at the age of twenty-nine he had a flourishing medical practice in Baton Rouge, Louisiana, as well as a wife and three-month-old son to grace a promising life. He was not overly involved in politics.

In addition, he gave no indication on the day of the assassination to anyone who knew him that he was about to do something so dramatic—

and certainly suicidal, as well as humiliating for his family. He'd been on a family outing and had just ordered new furniture for his home. Yet despite these nagging questions, this case has been closed with the popular conviction, spurred by police efforts to close the file and by a popular enthusiasm to follow the police lead, that Weiss was indeed an assassin. Yet was he?

A countertheory has it that Weiss argued with Huey Long in the shadowy corridor of the high-rise statehouse and struck him in the mouth. In defense or retaliation, Long's bodyguards then shot at Weiss and, in the fusillade, accidentally hit Long with a shot that proved fatal. It is a fact that Long had a cut on his lip that had occurred that evening. He told a nurse that he had been struck. If not by Weiss, then by whom? And if by Weiss, then why was Weiss gunned down rather than merely grabbed and pulled away? When one looks hard at this incident, the questions are both intriguing and ominous, and not all historians are satisfied that this tragic tale is a closed chapter.

Indeed, when *American Heritage* magazine asked noted Civil War documentarian Ken Burns in 1990 which mystery in American history he would most like to see resolved, he said: "More than anything, I would like to know who really shot Huey Long or what actually happened in the marble back corridor of the Louisiana State Capitol on September 8, 1935."

After looking closely and intently into the facts, I was of the same mind.

MY INVOLVEMENT BEGAN IN THE MID-1980S. I HAD JUST WRITTEN an article on the work and conclusions of three firearms experts, George Wilson, Marshall Robinson, and Larry Paul, on a puzzling Massachusetts case from the 1920s in which two men, whom witnesses described as "Italian looking," robbed and gunned down, in daylight, a paymaster, Parmenter, and his guard, Berardelli, who were delivering a payroll in South Braintree, Massachusetts. Investigators at the murder scene recovered six ejected shell casings and traced them to three manufacturers: Remington, Winchester, and Peters. They arrested two Italian laborers who were members of a radical anarchist group, Nicola Sacco, twenty-nine, and Bartolomeo Vanzetti, thirty-two. Both men were in possession of pistols and Sacco's was the right caliber—a .32-caliber Colt automatic—to be one of the murder weapons. Sacco also had two dozen bullets on his person, some of which had been made by each of the three manufacturers of the crime-scene shells.

At their joint trials for the two murders, experts argued over whether the four bullets removed from the victims could be matched to Sacco's pistol. Their opinions had no basis in science as it was then, and one expert was even exposed as an outright fraud (a pharmacist, playing firearms expert, from upstate New York). In spite of all and in the anarchist frenzy of the time, Sacco and Vanzetti were convicted of murder and sentenced to be executed.

During their appeals, a committee hired Calvin Goddard, who had worked at the Bureau of Forensic Ballistics in New York, to reexamine the evidence with his newly devised comparison microscope. In the presence of one of the defense experts, he fired a bullet from Sacco's gun into a wad of cotton and then placed the ejected casing on the comparison microscope. Next to it, one at a time, he examined the collected evidentiary casings. In terms of the marks left by the gun, the first two casings were no match, but the third one was. Even the defense expert conceded the remarkable similarity between the casing from the crime scene and the one shot from Sacco's gun. That same year, 1927, the two defendants went to the electric chair. Vanzetti still proclaimed his innocence, while Sacco cried, "Long live anarchy!"

Through the years, opponents of the verdict have claimed that because the defendants were anarchists, their conviction was politically motivated and otherwise not based in solid evidence. Yet a subsequent investigation with better ballistics technology in 1961 supported Goddard's findings. Even so, in 1977, Governor Dukakis of Massachusetts issued a proclamation declaring the innocence of the two men, and "the case that would not die" regained its controversial status.

Wilson, Robinson, and Paul, as three of the most prominent firearms experts in the country, were called to reexamine the firearms evidence in the Sacco and Vanzetti case for the state of Massachusetts. The evidence they had to examine was, strangely but commonly, found in the personal and private possession of the son of the state's principal firearms expert. He had, as was the knife in the Alfred Packer case, cozened the items as a family heirloom rather than a historical artifact to which he had no rightful claim.

I was given the opportunity to review the work of these competent firearms experts for the purpose of writing an article about it and their findings for the *Journal of Forensic Sciences*. I did so and published a peer-reviewed article in the *JFS* in two parts in two successive issues of the journal.

When my involvement in the firearms reevaluation ended with the publication of my papers, I asked George Wilson: "Where do I go next? Is there another case like this to work on?"

His response was immediate. "The assassination of Huey Long is right up your alley," he said. "There's a very close question of accountability in that incident, and it is chock-full of firearms issues, which I know catch your fancy."

Wilson's comment inspired me to look into the history of this incident, and I soon agreed that there was merit in launching an investigation.

It is not possible to understand this knotty saga without background on Huey P. Long, Dr. Carl Weiss, and the state of Louisiana politics during the 1930s.

HUEY PIERCE LONG, BORN IN 1894, TOOK HIS NICKNAME, "THE Kingfish," from a character in the popular *Amos 'n' Andy* radio show. Many people viewed him as an arrogant, manipulative lawyer and politician who rewarded cronies and harassed those who opposed him. His manner of governing paralleled the methods of fascism, and he employed the Louisiana National Guard to intimidate his enemies. Within an already corrupt system of patronage, bribery, and extortion, he thrived and eventually became the state's governor. Yet as much as he was feared and hated, he was also hailed by many as a charismatic populist who provided free textbooks to poor children, built bridges over major rivers, and paved the state's dirt roads. His political success was based largely on his reputed empathy for the exploited and underpaid white working class.

In 1930, Governor Long became a United States senator (defying the state's constitution by holding two powerful political offices at once).

Showing disdain for Franklin D. Roosevelt's programs, he prepared himself for a presidential bid. To encourage national support, Long organized "Share the Wealth" clubs and authored books to advance his cause, such as *Every Man a King*. He called for the common people to bring an end to the unrestricted power of the corporate rich and to demand better benefits and working conditions by redistributing the nation's wealth.

Yet the more popular he grew with the masses, the more those he threatened despised him. It was no great leap to predict that someone would try to eliminate him, and as a precaution he surrounded himself with bodyguards, known affectionately by him as his Cossacks. However, when the feared assassin finally arrived, he proved to be an enigma and possibly not the actual assassin at all.

Carl Weiss was born into a German Roman Catholic family in Baton Rouge in 1906, and he received his medical training from Tulane Medical School and Turo Infirmary in New Orleans, as well as at The American University in Paris, France. By 1935, he was a well-known ear, nose, and throat surgeon in the state. He had married into a powerful Roman Catholic family from Opelousas, Louisiana. His father-in-law, Judge Benjamin Pavy, was an outspoken critic of Huey Long; Long reportedly repaid the judge's effrontery by having two of Pavy's daughters dismissed from their teaching posts. Some sources say that Long had also threatened to "tar-brush" the family by leaking a rumor about African-Americans in their ancestry.

Dr. Weiss was a sensitive and scholarly man whose pictures portray him in that light. Perhaps he took the insults to his father-in-law personally or felt inclined to defend his wife's family. At any rate, on Sunday, September 8, 1935, he went into the state capitol in Baton Rouge armed with a 1910-model .32-caliber Fabrique Nationale semiautomatic pistol and dressed in his Sunday-best white linen suit.

It was just after nine in the evening. Senator Long was politicking in a late-night session of the Louisiana legislature. Dr. Weiss had left his home, situated just beyond a hefty stone's throw from the statehouse, to make house calls in his professional capacity as a physician. He parked outside the statehouse and entered. Long and Weiss met face-to-face, probably for the first time, in the north corridor of the building's second floor. The incident occurred near the anteroom to the governor's office just after Senator

Long had exited the House chambers at the east end of the corridor. With him were State Supreme Court Justice John Fournet and Long's usual retinue of seven bodyguards, or "Cossacks," as their man-in-charge, Joe Bates, termed them. So much is fact and up to this stage a matter of little dispute. What happened next has been mired in doubt, controversy, and confusion.

That a shoot-out of monstrous proportions occurred there was plain to the eye of any beholder. That there were any number of eyewitnesses, both near and far, at the moment of the shoot-out was beyond cavil. That Dr. Weiss was riddled with bullets, leaving in his body some sixty-one entry and exit holes, was established in the quieter aftermath of the fusillade. That Senator Long was struck in his abdomen by at least one bullet was indisputable, and the senator himself stated as much. He also bore a cut lip from the fracas.

Some say Weiss fired first, hit Long, and was met with a defensive hail of bullets. Others say that he never shot his gun but just punched Long in the mouth, cutting his lip. One witness stated that it had misfired.

Huey Long survived for some thirty hours with his wounds, inflicted by either one or two bullets (depending on which story is true) that had been fired during the melee in the capitol corridor. One bullet was to the abdomen, another reportedly to the lower spine. When a nurse asked Long about the abrasion on his lip, he supposedly replied, "That's where he hit me"—apparently referring to Weiss. He sustained massive internal hemorrhaging after bungled surgery at the Our Lady of the Lake Hospital in Baton Rouge—the surgeon failed to notice until too late that a bullet had nicked one of Long's kidneys. Senator Long quickly weakened, and on September 10, 1935, he was pronounced dead.

There was a cursory investigation into what had actually occurred, which involved crime-scene sketches and a hasty inquest, eight days delayed. The official story, printed in the newspapers, stated that Weiss had fired the fatal bullet and he was killed immediately. Years after the incident, Judge John Fournet described for an interviewer a strange expression on Weiss's face, though he did not elaborate on what he meant. He also said that a small black gun went off within a foot or two of him, but he did not say who was holding it. (Since he was discussing Weiss in this interview, it is assumed that he meant that Weiss was holding the gun.)

Fournet added that bodyguard Murphy Roden grabbed Weiss's gun as it went off, and the bullet struck Long in the right side, at a downward angle.

A journalist who was down the hall at the time said that all of the bodyguards joined in the shooting and that the man who had shot Senator Long pitched forward and died. The narrow, marble-lined hallway was filled with smoke from the shooting.

An unpublished report held that one of Long's own bodyguards turned on his benefactor and gunned him down, planting Weiss's weapon on him to frame him as the culprit. Adding to the controversy, a rumor had it that one of Long's bodyguards mumbled in a drunken state that he had killed his "best friend" and was immediately hustled out of state. Was Weiss just a patsy for traitorous or reckless bodyguards?

Weiss's family was resolute that he was no assassin. They refused the media's version of events, which they considered offensive and based in lies. It is a fact that no credible motive was ever unearthed for Weiss to have done this deed.

Despite many questions and much skepticism about the official version of the incident, neither body was autopsied with a full internal examination. And whatever physical evidence was collected at the scene of the shooting was concealed from the public and, over the years, disappeared due to a lack of documentation. Even worse, the East Baton Rouge DA stated that Long's associates had pressured him not to hold an inquest. Baton Rouge coroner Thomas Bird had wanted to do so, but the bodyguards had resisted, so it did not take place until September 16, eight days after the incident. By this time, all the participants had been thoroughly debriefed by General Louis F. Guerre, superintendent of the state's Bureau of Investigation and a Long crony, who ran a one-man investigation that resulted in a report to which only he had access, another item of investigative value that has vanished into the mists enveloping the Louisiana bayous.

WHO SHOT WHOM AND HOW MANY SHOTS WERE FIRED BY EACH of the participants only starts the query ball rolling. On Weiss's side, we have the following unanswered questions:

- Would a twenty-nine-year-old professional person of little or no political leanings, with a blossoming career as a physician, highly regarded in the community, of Louisiana-rooted German Catholic stock, and with a wife and infant son, throw all of this to perdition in a certain-to-be-suicidal assassination effort?
- Would an assassin doomed to die at 9:20 P.M. have placed a call only an hour earlier, at 8:15 P.M., to the home of another physician, set to assist him in an operation the very next morning, to alert him to a change of hospital location for the operation?
- Would a man whose mind was intent upon an assassination have lolled away the last day of his life in gaiety with his family, eating, joking, swimming, and generally acting as if life was good and was to go on for him?
- Intending to assassinate Long, would he have driven his car from his home, which was so close to the capitol, and parked in front of the capitol? So close to home, why did he not walk to the capitol?
- Was Dr. Weiss, as Mark Twain satirically described every person, "like a moon with a dark side which he conceals, if he can"?

Yet on the other side of the coin, if Dr. Weiss did not hover in the marble-pillared shadows of the capitol's marble-lined corridor with a fixed design to kill Senator Long, why was he there at that late hour? If his presence was but an unannounced visit to a legislator related to his wife, as some have said, why was he observed to be acting covertly in the corridor?

Were he bent on assassination, some analysts have argued, he would have shot Long in the head, not the abdomen. Others have indicated that Weiss had wanted to see Long on a political matter, had reacted in anger by hitting him with his fist, and had paid for this rashness with an instant fusillade from the bodyguards, one of whose bullets accidentally caught Long himself. So then we must ask whether the bodyguards appointed as guardians of the senator's person overreacted to the perceived danger of the presence of Dr. Weiss.

Another consideration is a theory that Dr. Weiss had been affected by some toxic substance, which he may have ingested or injected, or that he

possibly had a brain tumor or some neurological condition that had pro-
voked this bizarre and irrational behavior.

As I looked into the matter, the questions were plentiful, puzzling, and
provocative, with the answers appearing to be sparse, speculative, and spo-
radic. For over fifty years, those who had studiously sought the answers,
from T. Harry Williams to Hermann Deutsch to David Zinman to Ed
Reed (to mention only those who have written books or parts of books
devoted to the mysteries of the matter), have relied on the statements of
the men who survived the debacle in the Baton Rouge capitol corridor, as
well as others who could provide oral reports. No one had thus far given
conscientious or even more than passing attention to a scientific analysis of
the physical evidence that such an incident had to have generated.

I wondered why.

Yet to pursue this case further, I had to find members of the surviving
Weiss family and ascertain their views on any further investigations by me.
Thomas Weiss, Carl's younger brother (whom everyone calls Tom Ed),
was nineteen years old at the time of the incident. I learned that he was on
the medical staff at the Oxner Institute in New Orleans, and a highly regarded rheumatologist. Since I was about to attend the American Academy
of Forensic Sciences meeting in that city in 1985, I used the opportunity to
invite him to lunch.

He agreed to meet me, and he turned out to be passionately Old School
Louisiana—a complete gentleman and a devout Roman Catholic. At lunch
at one of the area's fine restaurants, I presented my concerns about the case
and suggested that this was a prime occasion to put modern science to work
on the unanswered but possibly not unanswerable enigmas surrounding this
volatile matter. He was interested, so I explained that I understood that on
the day before Carl Weiss died, the Weiss family had gone out shooting at
trees along the river. Tom Ed acknowledged that this was correct. So I said,
"We might be able to get a bullet out of a tree to compare with the bullet
and gun that killed Huey Long to see if there's a connection."

He seemed to take my observations to heart. I asked him about Carl
Weiss, Jr., who had been only three months old at the time and whose
mother—Carl Weiss's widow—had left the state shortly after she lost her

husband. Would he be willing to back my investigations? Tom Ed shook his head and told me, "He's not ready."

Disappointed, I let it go for the time being, but a year later, Tom Ed called me and said, "My nephew is ready. He'll talk to you now."

I made arrangements for an overnight visit with Carl Weiss, Jr., in Garden City, New York, where he was a practicing orthopedic surgeon. Upon my arrival, he picked me up at the train station and said, "You're going with me to the hospital. I'm on call tonight and there's been an automobile accident." I agreed to go along and wait for him. But it wasn't going to be that easy.

"No, no," he said. "You're going in with me."

He went into the prep room and stripped down to prepare for surgery, indicating that I should do the same. So I did, and I walked with him into the surgical suite. He introduced me to the others in the room as a doctor from Washington, D.C., and I did my best to play along.

Yet for me, it was a surreal environment. Loud music played in the background while he conversed with the anesthesiologist. The accident victim, a woman, lay there on the gurney with a compound fracture. A bone was sticking out of her arm. I wanted to stay in the background and not get too close, but Dr. Weiss called me over to look. Then he got his scalpel ready for the inevitable incision. As the blood began to flow, that was it. I couldn't make it. I had to leave.

I excused myself and went outside to sit in a chair with my head between my knees. His chief nurse came out, took one look at me, and said, "You're no doctor from D.C."

I just looked at her sheepishly and admitted, "You've got my number."

Obviously, Dr. Weiss was putting me to the test.

Afterward, we went back to his house for dinner, and everything was diplomatic and decorous with all references to the surgery off the table for the moment. But then another test confronted me. During the meal, he invited me to try some white wine. I looked at the label and saw that it was Westchester County wine. Hesitant, I had a glass or two, but it was truly unpalatable. As dinner neared its end, Dr. Weiss asked, "How do you like that wine?"

Now I was on the spot. I had to make a quick decision: Do I lie or do I tell the truth? So I looked him squarely in the eye and said, "It's not to my liking."

With that, his wife piped up, "I like that man! He tells the truth."

Dr. Weiss looked at me unreprovingly and indicated that if I was OK with his wife, I was OK with him.

That was the opening wedge for a further investigation into the death of his father and, quite possibly, an exhumation of his remains. After that exchange, the tenor of everything changed. Dr. Weiss showed me through his study upstairs and explained that he hadn't been very interested in his father's death, but he'd help if he could. He left me alone to look through several pages of a diary that his father had written as he'd traveled through Europe. The son had kept one, too, when he'd made a similar trip many years later.

As I read the diary entries, I noticed that both of them had read *All's Quiet on the Western Front,* and both had written their own reviews of it. It was crystal clear to me then that Dr. Weiss, the son, had actually been quite deeply and disturbingly touched by his father's death, even to the point of reliving step-by-step his father's European trip. As I read, I could see from the poetic entries in his diary that Carl Weiss, Sr., had been a caring and sensitive, even a soulful, person.

Truthfully, I could imagine him painfully suffering under the political bullying in the public forum of Huey P. Long. I could also see that his anguish over Long's tirades and antics could have reached the breaking point, he being a gentle, scholarly, and even slightly withdrawn professional. The question wasn't, however, whether it could but whether it did.

As my random thoughts drifted back to the present, I reminded myself that my investigations were still aborning. The next day, I entrained back to Washington, D.C., confident in the knowledge that the immediate family of Carl Weiss was solidly behind my continuing investigations. Back in the utter clutter of my office, I drafted formal legal consents for both Tom Ed and Carl Jr. to sign. In my cover letter to Carl Jr., I took the liberty of calling upon St. Paul to signify that an exhumation need not be miraculous for new facts to emerge. As I put it, "It's not only on the road to Damascus that revelations are made." In returning the signed consents, he remarked that the reference to St. Paul had struck a responsive chord with him.

IT WAS NOW TIME TO SEEK ABOVEGROUND FOR THOSE ITEMS OF physical evidence to which I could apply modern scientific methods, as had been done with Sacco and Vanzetti and with the Alfred Packer case. But I was in for a rude surprise. It turned out that the reasons for inattention to the physical evidence over the past half-century were not at all obscure:

- No autopsies were conducted on the bodies of either Senator Long or Dr. Weiss, so there were no medical records or findings.
- All of the many bullets that had entered but did not exit Dr. Weiss's body were buried with him, save for one .38- and one .45-caliber bullet, which were recovered from his remains when "they were just found under [his] skin" by East Baton Rouge coroner Thomas B. Bird. These bullets had long since disappeared and were untraceable at this late hour.
- The bullet (or bullets) recovered in the hospital and/or mortuary from Senator Long's body has long been among the missing items of physical evidence that could contribute to unraveling this historical puzzler. (It was never proved how many bullets had actually been removed from Long or whether any had been from a .32-caliber gun of the type owned and carried by Dr. Weiss in a gun sock in his car.)
- All the many spent cartridge casings and the bullets that had missed their human targets in the capitol corridor had literally been swept under an investigative rug from which they had never reappeared.
- General Guerre's report on the investigation, and the police sketches and photographs of the crime scene, were missing.
- Most important, Carl Weiss's pistol was not to be found.

Clearly, my way was not going to be a yellow brick road, but believing firmly that no physical evidence is ever lost, just temporarily unavailable, I was determined to squeeze the historical record for every ounce of information I could eke out of it.

In pursuit of this goal, in 1986 I wrote to then Senator Russell B. Long, the son of Senator Huey Long, to obtain his assistance. He replied on June 27, 1986, that his views could be found in the *Congressional Record*. He was convinced that he knew who had shot his father and he could "see no point in encouraging the present investigation," even though I later learned that he was then privy to information about the location of at least the .32-caliber pistol that had belonged to Dr. Weiss. The mysterious owner had offered to sell it to him, but he did not provide that information to me.

Other sources would have to stand in for Senator Long. The first place to look for the most responsible portrayal of controverted past events in an incident in which homicide is suspected is the record of testimony given before a judicial or quasijudicial body. A coroner's inquest is a proceeding of a quasijudicial nature convened to assess the possibility that a death has resulted from criminal or similar suspicious causes warranting further investigative or prosecutorial action.

In the case of the separate deaths of Carl Austin Weiss, M.D., on September 8, 1935, and Senator Huey P. Long, Jr., on September 10, 1935, two inquests were initiated by Dr. Thomas B. Bird, the coroner of East Baton Rouge Parish. The inquest into the death of Dr. Weiss, which commenced on September 9, was adjourned to September 16, 1935, at which time sworn testimony was given.

The reason or reasons for the delayed hearing could be attributable to the fact that the witnesses were not immediately available or, more suspiciously, because of the need for extra time for the witnesses to harmonize their separate recollections of the occurrence. The inquest into the death of Senator Long was of shorter duration, taking place informally at the Rabenhorst Funeral Home in Baton Rouge. No testimony was taken, it being simply an external observation of the remains of Senator Long.

The nearest one can come to an "official" version of those pivotal events on September 8, 1935, is that contained in the transcript of the coroner's inquest on the death of Dr. Weiss. However, there are those who consider this transcript to be less than an authentic account, since its reporter,

Glenn S. Darsey, inexplicably did not swear to it as a true and correct transcript until May 5, 1949, some fourteen years after it was taken down.

Moreover, no actual transcript in official and original form of the Weiss inquest exists in any court, police, coroner, or other official file in the state of Louisiana. The Darsey-endorsed transcript became a matter of public record only on September 10, 1985, the fiftieth anniversary of the death of Senator Long, when Senator Russell Long entered it into the *Congressional Record*—as he had informed me. The seventy-two original pages of the transcript have now been compressed into the three-column format of ten pages.

Unlike this transcript, the verdict of the coroner's jury is a matter of public record in the court files in East Baton Rouge Parish. In the case of Dr. Weiss, the jurors found that his death was the result of "pistol wounds of head, chest and abdomen [homicidal]." Senator Long's death was declared to be due to a "pistol wound of abdomen [homicidal]," with no culprit identified in either case or any determination as to whether the homicide was criminal in nature. By Louisiana statute, Coroner Bird was supposed to issue a verdict assigning guilt, but this was not done.

There were at least three pivotal matters that I hoped the contents of this inquest transcript would resolve: How many shots did Carl Weiss fire? Did he fire the shots as an aggressor or in self-defense to the Cossacks shooting at him? And was he shot in the back while downed in the corridor, lying defenseless and disabled?

Yet on the point of the number of shots fired by Dr. Weiss, there is irreconcilable conflict in the testimony of the witnesses at the inquest. Some said that Dr. Weiss fired one shot. Others thought he got off two shots, and no one confidently stated that he discharged more than two shots.

Murphy Roden, a principal bodyguard for Huey P. Long with a reputation as a bulldog of a policeman and a central figure in my investigation, was of the opinion that only one shot was fired from Weiss's gun. Elliott Coleman, another bodyguard, thought Weiss's gun had fired "when the second shot was fired." So whether Dr. Weiss got off the first or second shot fired that evening was also left for posterity to ponder.

Bodyguard Paul Voitier mentioned Weiss having shot but once, and Ad Riddle was of a similar mind. But Judge Fournet, in testimony that could only be charitably described as in complete disarray, was of the opinion that Senator Long "was shot twice," whether by Weiss or someone else he did not say.

The mystery of how many shots, if any, were fired by Dr. Weiss perceptibly deepens when it is observed that the FBI's Freedom of Information files in Washington, D.C., bear evidence that, potentially at least, *three shots* might have been fired from Dr. Weiss's pistol. The FBI got involved, but never embroiled, in the post-tragedy investigation when General Guerre wrote to J. Edgar Hoover asking his assistance in tracing the .32-caliber Fabrique Nationale (Browning design) semiautomatic that was said to have been confiscated from Dr. Weiss. A picture of this weapon was attached to Guerre's letter to the FBI, and I photocopied it from the FOIA files, first disclosed at the request of Ed Reed, a Louisiana author and a most informed commentator on the killing of Huey Long.

The photograph shows the weapon and an empty magazine adjacent to it. Scattered about the magazine and the weapon are five cartridges. Apparently, these five cartridges were all that remained unfired of the magazine load that Dr. Weiss had brought with him to his confrontation with Senator Long.

Yet this photograph gives one the willies, for the top three unfired rounds displayed in it are not genuine rounds. They are cutouts of rounds that have been pasted in and rephotographed with the Weiss weapon. Thanks go to my photographer, Jim Kendrick, for this insight. The whys of this fabrication were unacknowledged and unexplained in General Guerre's cover letter. Is it too much to say that the photograph was "doctored"? If so, for what purpose?

Looking beyond that to what we know about the gun itself, the .32-caliber handgun displayed in the photograph submitted to the FBI was a 1910 manufacture fitted with a magazine that could accommodate a maximum of seven rounds. Therefore, if Dr. Weiss carried the weapon with a fully loaded magazine into the capitol and five unspent cartridges were found in the magazine subsequent to the shooting, then it must be as plain as Nancy Drew that two shots were fired from that weapon.

Yet, as the old song teaches, "it ain't necessarily so." I propose an alternative explanation.

FIRST, WE MUST CONSIDER THAT ALL THE SIGNS IN THE TESTIMONY at the coroner's inquest pointed to Dr. Weiss's having fired his weapon quickly, without waiting to adjust his sights. Judge Fournet recalled "a man dressed in a white suit" who "flashed among us. He moved hurriedly . . . flashed a gun and shot almost simultaneously." Murphy Roden's testimony was, in this regard, entirely in accord with Judge Fournet's recollection. "He brushed right through," remembered Roden, "pulled a gun and fired at Senator Long." Lickety-split and the deed was done. No hesitation, no wasted motion, just a few steps and a shooting.

However, three witnesses spoke of Dr. Weiss's left hand being in action as well as his right, which held the gun. Paul Voitier saw Weiss's left hand "covering the weapon he held in his right hand." Ad Riddle thought Weiss was holding the gun in "both hands." Judge Fournet noticed that Weiss had the pistol "in his right hand trying to shoot it with both hands." But at that moment Weiss had already gotten off one shot, in Fournet's opinion, when he was seen to be "in a crouching position" seeking to ward off the cannonade of fire from Senator Long's "skull-crushers," as the body-guards have been disparagingly named.

What, if anything, might be the significance of the testimony of Voitier and Riddle that Weiss's left hand was positioned over or about the weapon he held in his right hand? There are two reasonable hypotheses. Either Weiss may have been using his left hand to steady the weapon, which would locate his left hand under the weapon for this purpose, or he may have been in the process of pulling back on the slide in order to chamber a round so as to make the weapon ready for immediate firing. Any lay observer could recognize the credibility of the first hypothesis, but only one with a basic understanding of firearms would perceive the likelihood of the latter hypothesis.

Assuming that this weapon carried a full magazine load of seven cartridges, a simple pulling of the trigger would not cause a round to fire. The weapon is a semiautomatic, enabling the automatic chambering of a round once a prior round has been fired. But the firing of the first round will not occur until a round is manually chambered by pulling the slide rearward, with the magazine inserted, and then releasing the slide.

Recognizing that there would be a time lag involved in this manual chambering procedure—minuscule though that time lapse might be—a person fixedly bent on shooting such a weapon without the flicker of an eye's intervention would seem inclined to chamber a round in advance of the actual occasion for the shooting. With split-second timing an all-important consideration and no time willingly left to fate, such a precautionary advance chambering would be de rigueur. Now let's apply this understanding to the events in the Louisiana state capitol on September 8, 1935.

If Dr. Weiss was determined to kill Senator Long, and had a preconceived design to do so, then one could hypothesize that he would have a weapon ready for action as he lurked in the corridor of the state capitol waiting for Senator Long to come into view. To be ready for action, a round would have to be chambered, enabling the firing of the cartridge simply and instantly upon pointing the weapon and pulling the trigger. Yet as we have seen, two witnesses at the inquest distinctly recalled Dr. Weiss's left hand being in a position reasonably interpretable as serving to chamber a round. If so, then Dr. Weiss was either foolhardy in being unprepared for his criminal enterprise or he did not have homicidal intent when he entered the capitol and waited in the corridor.

The number of cartridges displayed in the photograph that General Guerre submitted to the FBI is also instructive. The maximum number of rounds that the magazine for the 1910-model .32-caliber Fabrique Nationale would hold is seven, but if the weapon had been "combat loaded," then the total number of cartridges it could hold would be eight. A "combat-loaded" weapon is one that has a round chambered and a magazine full to its maximum capacity.

Once again hypothesizing a predetermined murderous plan by Dr. Weiss, it would be logical to suppose that the .32-caliber weapon would have been cartridge-full, to its "combat-loaded" capacity of eight rounds. Yet five rounds are pictured as not having been fired, leaving three as having been discharged—at the maximum. The testimony at the coroner's inquest, however, leaves no room even for speculation that Dr. Weiss fired three times; two was the maximum number of shots that the witnesses who could recall the events reasonably assigned to his shooting.

Assuming that Dr. Weiss fired only two shots, then either he was not combat-ready—that is, not prepared to do deliberate mayhem or murder with a fully loaded eight-cartridge weapon—or he had a round chambered and one less than capacity in the otherwise seven-round-capacity magazine. A careful planner of a murderous venture would not seem to go into a catastrophic shoot-out with one cartridge less than the total he could use or might need. The five remaining unfired rounds, as depicted in the picture submitted to the FBI, together with the testimony at the coroner's inquest, give captivating strength to one probable hypothesis: Dr. Weiss did not enter or remain in the corridor of the state capitol with a mind predisposed to murderous foul play.

And there is more in support of this hypothesis: the proven methodical nature of Dr. Weiss in his other life's activities. His meticulous character stands out boldly in the testimony at the inquest of Dr. J. Webb McGehee, who asserted that at 8:15 P.M. on the day of the tragedy, Dr. Weiss placed a call to him to confirm that he knew that an operation by Weiss scheduled for the following morning had been shifted from Our Lady of the Lake Hospital to the Baton Rouge General Hospital.

Would a man whose mind was concentrated on a surgery he was to perform the very next morning purposely go into a shooting match with Senator Long's "Cossacks," with a weapon that could not be fired without expending precious time in chambering a round? A man like Dr. Weiss, with an overweeningly punctilious frame of mind, would hardly conjure up a murderous design and neglect to have a round chambered that would ensure the smooth and certain execution of his plan. Yet the testimony of Riddle and Voitier, as well as the existence of the five unexpended cartridges, lends persuasive weight to the possibility that Dr. Weiss was designedly ill-prepared to execute a murderous assault.

A further note on the operation of semiautomatics in general points to the lack of chambering of an initial round as well-nigh dispositive proof of Dr. Weiss's intentions in loitering in the corridor of the Louisiana state capitol. Semiautomatics, unlike revolvers, are well-known to malfunction when a cartridge jams during the chambering process. It is the recognition of this dire reality that for a long time kept the police in determined possession of revolvers rather than semiautomatics.

Knowledgeable owners of semiautomatics are presumed to be aware of this operational mischance, which could make the difference between a split-second and a delayed firing. Dr. Weiss would hardly be expected to rely on the surefire chambering of a round in his .32 semiautomatic at a moment's notice when he could have easily avoided any operational mishap by chambering a round in advance. Yet the testimony of Riddle and Voitier, the finding of the five unexpected rounds, and the testimony that he got off two shots all give sturdy support to the hypothesis that Dr. Weiss did not enter the affray with a round chambered in his weapon.

Thus, he was either a monstrously careless murderer or not a murderer at all, according to the logic of firearms scholarship.

ANOTHER QUESTION THAT INTRIGUED ME ARISING FROM THE coroner's inquest concerned the type of weaponry that was used against Dr. Weiss. With undeviating certainty, the testimony of witnesses was to the effect that no machine guns were used to defend Senator Long. The firing from the bodyguards was variously described as a "riot" of firing or a "whole volley." Whatever the language, the firing was said to be loud, intense, and even explosive in nature, but machine guns were said to be nowhere in evidence in causing the din.

C. Sidney Frederick, the district attorney of the St. Tammany–Washington District, was in the capitol corridor at the time of the barrage of shooting. He testified that he "saw revolvers in the hands of more than one person" but no machine guns. Another witness, one Earl Straughan, whose business in the capitol on September 8, 1935, was not stated, saw four or five men with "six-shooters, like the City Police carry here," but "not automatics."

In direct contrast to this testimony, Murphy Roden testified that he had wielded and fired a ".38-caliber Colt super-automatic." Indeed, none of the bodyguards spoke one word about their firing or their observing the firing of any revolvers. Roden said he'd emptied his semiautomatic of its ten rounds. Joe Messina stated that he had also "unloaded" his weapon at Dr. Weiss, but he did not specify the model of the weapon or the number of its cartridges. Elliott Coleman, another Cossack, had gotten off three shots at Dr. Weiss before his attention was directed elsewhere. Paul Voitier admitted having fired five times at Dr. Weiss with his .38-caliber handgun. He was not asked to state, nor did he volunteer, whether it was a revolver or a semiautomatic, nor of what manufacture.

In all, seven of Senator Long's bodyguards, all of whom were employees of the state's Bureau of Criminal Identification, were summoned to testify at the coroner's inquest. Only George McQuiston refused to testify, and he was excused before he was called upon to explain the basis for his refusal. Of the others, only Roden, Coleman, Voitier, and Messina gave evidence of their having fired at Dr. Weiss. Both Joe Bates and Louis Heard specifically denied having fired at all, and Bates insisted, "My gun was never pulled from my pocket."

Only Murphy Roden and Paul Voitier gave any particulars at the 1935 inquest on the weapons they were carrying or using in the affray. Roden's recollections were specific and detailed. He had a ".38-caliber Colt super-automatic" both in his possession and in action against Dr. Weiss. In later years, Roden explained that his pistol was a "Colt .38 special on a .45 frame." The .38-caliber Colt superautomatic has been touted as "an arm of unsurpassed power and efficiency" that "will stop any animal on the American continent." If there is even a half-truth in those high-blown commercials, the weapon would still suffice to immobilize a 5-foot-9-inch, 132-pound person like Dr. Weiss—that is, if Roden's aim was true.

Author David H. Zinman tells us that Roden was a "sharpshooting cop from Arcadia, Louisiana." The story is told of him going to Washington, D.C., by car, starting out at the same time that Long left by train and driving literally like a maniac to arrive just before Long. As the kingpin among the Cossacks, the man was extraordinary and was the most likely person to have shot Long by accident. Since his personality was aggressive, he was probably the first in the melee to pull a gun. He had fired quite a number of times. Roden was reportedly so sure-eyed that he "could empty a pistol into a four-inch target at 50 feet." It is also known that at an FBI training course in 1937, Roden ranked first among his classmates in handgun marksmanship.

In that case, it would have been no great feat for him to have put into Dr. Weiss all ten rounds he said he fired from what must have been a fully loaded .38 Colt (which the 1940 *Shooter's Bible* lists as having a nine-cartridge magazine, and one chambered). The reality of such a probability is not diminished by Roden's testimony that he "wouldn't swear any one of my bullets hit him." Standing "right on him," as Roden conceded he was, it would have taken a blind man shooting with his off hand with an unloaded pistol to have missed the target.

Paul Voitier's testimony concerning the weapon he fired was limited to affirming that he shot a ".38." Neither Coleman nor Messina specified the make, model, or type of weapon they possessed or fired, but twenty-five years later Coleman is reported to have said "Those boys all had .44's and .45's." Author Ed Reed tells us that "Joe Messina carried a .45," but he neglects to mention the basis for his knowledge.

A reporter by the name of Ewing supposedly went over to the body of Weiss, which after the shoot-out was "lying on [its] face at a forty-five-degree angle against a south wall of the corridor." Ewing tells us, through Zinman's reportage, that the "bodyguards were milling around as if demented, with Colt .45's in their hands." Whether Ewing took any particular care in determining the caliber and manufacturer of the guards' weapons is unknown, but in light of the fact that Zinman, also a reporter, refers to the "crash of exploding revolvers," there is at least a reasonable doubt that Ewing was any more accurate in distinguishing revolvers from semiautomatics and .38-calibers from .45-calibers, et cetera. However, someone must have wielded and fired a .45, since Coroner Bird said that he removed one .45-caliber bullet from Dr. Weiss.

Deutsch has said that "most of the large-caliber cartridges [.44's and .45's] also carried hollow point bullets, which have a mushrooming effect." Roden has confirmed that his .38 was loaded with hollow-point ammunition.

If the skull-crushers, as Long's brother Julius termed the bodyguards, were using hollow-point ammunition, then there would be credible grounds for believing that Dr. Weiss's body suffered substantial injury as those bullets struck him and expanded as they penetrated his tissues. In the case of a hollow-point bullet, the temporary wound track through tissue is considerably larger, due to the expansion of the bullet's nose as it travels through bodily tissue.

But even if the bullet does not mushroom and increase in width, its augmented velocity (due to its larger load of propellant) will cause a correspondingly greater loss of kinetic energy while it penetrates the tissue, resulting in the likelihood of more dramatic bodily injuries. This added stopping-power of hollow-point ammunition would, as a consequence, have more speedily put an end to any actual or threatened use of force by Dr. Weiss.

Myths abound concerning the effects of hollow-point ammunition, and we can see their influence on the bodyguards' recollections. Yet these bullets do not tear a body apart, leaving it in the condition of "chopped meat," any more than any other projectile does. Thus, at autopsy it is not possible to conclude "that the individual was shot with a hollow-point rather than a solid lead bullet" unless the bullet itself is retrieved and evaluated. In addition, "wounds in the skin . . . are the same in appearance" for both hollow-point and other projectiles. This scientific recognition casts doubt on a portion of Roden's recollection of his encounter with Dr. Weiss.

According to author Hermann Deutsch, sometime in the early 1960s, contrary to what he had said at the inquest, Murphy Roden recalled that he had "fired one shot into [Dr. Weiss's] throat, under his chin, upward into his head and saw the flesh open up." The passage of twenty-five years had not dimmed Roden's memory of this startling event. In fact, it had measurably sharpened. Nothing appears in Roden's testimony under oath before Dr. Thomas Bird on September 16, 1935, about having shot Dr. Weiss through the throat at a near-contact range. Rather, on that occasion, Roden "wouldn't swear any one of my bullets hit him."

More to the present (hollow) point, however, the fact that Roden saw Weiss's "flesh open up" upon its being struck by his bullet suggests that the skin was torn apart and visually devastated upon the bullet's entry. Yet if Roden is correct about his shot, Dr. Weiss's skin should have borne evidence of a typical near-contact entrance wound—small, rounded, and abraded, with ample evidence of powder tattooing. The short range of fire and the "souped-up" .38-caliber superammunition would have made a smaller rather than a larger, opened-up wound of entry. So Roden's remembered account is problematic.

Another myth that has made the rounds concerning hollow-point bullets falsely teaches that hollow-point bullets fragment, or blow up, in the body. That even solid bullets will break up so long as they strike bone goes without saying, but there is nothing intrinsic to hollow-point bullets that will cause them to go to pieces in conditions in which solid bullets would not. But unlike solid bullets, which can ricochet or fragment upon hitting bone, hollow-point bullets are less likely to ricochet upon striking a hard object.

The mushrooming aspect of hollow-point bullets, together with the quick diminution of their kinetic energy in a body, gives greater assurance that once having penetrated Weiss, they would remain there—in contrast with the behavior of solid bullets of similar muzzle velocity. It has been described as one of the virtues of hollow-point bullets that they tend to stay in the body, making it unlikely that they will exit and injure bystanders.

Thus, we find that the coroner's inquest (the authenticity of which is already questionable) only yields more conflicts and mysteries. We can't tell from these disorganized and vague witness reports how many bullets Dr. Weiss fired or how many bullets (or whose bullets) hit Huey Long. We can't tell whether Dr. Weiss was aggressing against Huey Long or seeking to protect himself. We can't explain why an otherwise punctilious man would go on such a deadly mission unprepared. At this time, in 1991, fifty-six years later, it seemed that the factual void in this matter was strongly pressing toward the last resort: new, previously undiscovered information based on an exhumation.

INSTEAD OF RELYING ON MEMORIES, WHICH ARE PRONE TO ERROR, we can look to the examination of Dr. Weiss's remains for some idea about what happened in the shoot-out. The most specific details appear not in the testimony at the coroner's inquest but in the prefatory note by Coroner Bird. In that bit of discourse, he related the results of his examination of Dr. Weiss's remains, which was only external and certainly cursory.

Dr. Bird counted thirty bullet holes in the anterior and twenty-nine bullet holes in the posterior of Dr. Weiss. Two additional indications of bullet strikes were seen on Dr. Weiss's head. One was from a bullet that had penetrated below Dr. Weiss's left eye and exited through his left ear. Another bullet had grazed his face after passing through the tip of his nose, but was not described as having entered through his skull. Dr. Bird stated an inability to decide which of the bullet holes in Dr. Weiss's body were entrance wounds and which were exit: "There were so many in every direction." Whether he took radiographs to assist in his tracking of the path of the bullets is not mentioned. Two bullets were recovered, one a .38-caliber and the other a .45-caliber. Bird removed them from "just under the skin." He neglected to make their exact location in Dr. Weiss's body a matter of official record. Nor did he mention examining Dr. Weiss's arms and hands for evidence of bullet trauma inflicted when Dr. Weiss was being gunned down.

And I found another matter of serious neglect: the crime-scene photos.

Crime-scene photographs can provide a world of informative detail, which can facilitate post-event reconstruction of the scene and can assist in verifying the accuracy of the oral recollections of persons at the scene. Photographs stop the action and hold it in an unaltered state for the assessment of posterity—that is, if the photography is skillfully executed. Whereas the memories of witnesses will invariably fade and/or be remade over time, photographs will remain crisply demonstrative of the event and of the persons pictured.

The only documented post-event photograph of the scene of the shooting in the Louisiana state capitol is that which shows Dr. Weiss lying supine on the capitol corridor hallway. His arms are outstretched, his right hand resting against a nearby marble pillar and the top of his head adjacent to the bottom of the marbled wall. Blood is evident on his head and torso, as well as in splatters on the wall above his head.

The picture of a dead Weiss in the corridor appears in Hermann Deutsch's book. No caption appears with it to credit the photographer who took it (whom I might then have tracked down). Deutsch says only that the picture was taken by "the official photographer of the State Bureau of Identification."

I looked at it many times during the course of my investigation but initially saw nothing of significance. That's because I was looking at it as Watson, the sidekick of Sherlock Holmes, might have. I was seeing, as Holmes would remark, but I was not observing. Having read all of A. Conan

Doyle's Sherlock Holmes at least four times and being enamored of Holmes's scientific methods, I should have been more alert.

In "The Adventure of the Speckled Band," Holmes and Watson investigate the case of a woman murdered in her own locked bedroom, though there appears to have been no forced entry. In an adjoining bedroom, Holmes examines the seat of a wooden chair. He seems satisfied with what he has found but refrains from commentary. Watson remarks that Holmes has evidently seen more in these rooms than was visible to him. Holmes responds, "No, but I fancy that I have deduced a little more." In other words, they saw the same things, but Watson did not observe them in the informed and logical manner that Holmes had done.

As I looked at the photograph of Carl Weiss lying supine on the corridor floor, riddled with the Cossacks' bullets, I suddenly saw it as Holmes might have. I observed that there was evidence of blood in his left hand, and I could deduce what it meant. He probably had been shot through his wrist or through his palm. That indicated to me a defensive wound. He had more likely thrown up his arm to ward off the barrage of bullets coming at him before he fell to the floor and died.

Perhaps that meant that he had not shot his gun at all. Perhaps he had hit Long in the face with his fist and had been shot down on account of it by Long's Cossacks. One of their own bullets, flying wild or ricocheting off the marble walls, may then have killed their boss. Of course, although not immediately plausible, the shoot-out might have been the cover someone in the corridor needed to silence Long's political posturing once and for all. No one denies that Long had many, many enemies roiled to a fever pitch by his proclamations. After all, were not Julius Caesar and Jesse James assassinated by one (or more) of their own kind?

I looked into the mortician Merle Welsh's report on Huey Long. Welsh, when interviewed later by historian William Ivy Hair, supposedly stated that Huey Long's private physician entered his mortuary during his preparation of Long's remains for burial and removed a spent bullet out of Long's abdomen that was larger than a .32. It seemed the time was ripe for me to interview Welsh before committing to an exhumation, which in my lights should always be the last resort.

I managed to locate him, and he agreed to do a videotaped interview at his home. Yet as the interview progressed, it was quite clear to me that he had more knowledge of the event than he was about to disclose. I got as much as I could, but I left with the feeling that there was much more to the story.

Welsh had cradled Dr. Weiss in his arms in the corridor as photographs were taken by state officials. He had prepared both Long and Weiss for their respective burials. But he was now hemming and hawing about the event and its aftermath. The interview was unsatisfactory in terms of any new insights gleaned from it. Merle Welsh was, in a word, a hard sell.

The crime scene was next on my investigative agenda. Luke Haag, a criminalist and firearms specialist from Phoenix, Arizona, and a key member of my team, accompanied me to the infamous marbled corridor at the Louisiana state capitol to examine the place where the shooting had occurred. We found holes in the marble floors and walls that at first blush might have been attributable to gunshots, then placed orange cones on the holes in the floor for the purpose of taking photographs.

Then, looking elsewhere in the hallway, we found similar holes in the marble that the shoot-out, occurring a distance away, could not have impacted. This finding caused us to reconsider whether any of the holes had been made by bullets.

As we stood quite apart from the original site, a tour guide brought a group of schoolchildren into the corridor. She pontificated that the holes in the walls had been made by the bullets shot at Weiss in 1935. The children placed their fingers into the holes as if the holes were a sacred shrine that they and they alone could touch. I realized that these holes were nothing more than the marble exfoliating or decaying away over time. These were not bullet holes at all, for if they were, how did they come to be located in places in the corridor quite far removed from the scene of the shooting?

I subsequently learned that the original marble slabs that were in place at the time of the shooting had been replaced and were currently in storage. Those "bullet holes" proudly described by the tour guide were not authentic at all. I realized that a proper scientific reconstruction of the crime could occur only if those original slabs were brought back and set into place. It could be done, but at great expense and by someone with engineering expertise to spearhead such a project. It seemed to me at that time that, through our work on this project, we might generate the requisite public support to have this site reconstructed, if only for the sake of historical verisimilitude.

My investigations aboveground had been tantalizing but unproductive, leading me every which way. Several times I had found only more confusion than at the start, or a dead end. Lacking the necessary physical evidence and having many more questions than answers about the controversial events on September 8, 1935, I decided that the remains of Dr. Weiss might possess the key to unlocking the mystic uncertainties in this matter. The only unexamined evidence we actually knew existed, although unrevealed but meaningful, were the bullets that were still lodged in Weiss's body. In addition, an exhumation would most assuredly generate public attention, which could encourage people who possessed the missing items or unvoiced details to come forward. It was my purpose to utilize forensic science resources to give a factual base to the resolution of the many uncertainties surrounding the deaths of Long and Weiss. I discussed this with Tom Ed Weiss and Carl Weiss, Jr., and they agreed with me that an exhumation was the only remaining alternative for clarification of the event.

But who should be exhumed? Huey Long or Carl Weiss? In the case of Huey Long, I lacked and was unlikely to obtain the consent of his family for an exhumation. In addition, there were two practical barriers to his exhumation. He was buried under a monolithic monument standing in front of the capitol with a statue of him topping it off, a most unlikely situation for an exhumation. Further, he had been surgeried literally to death, masking or obliterating, as in the case of John Fitzgerald Kennedy, the presence of anything of forensic merit to bring the controversy to a halt.

When it was learned in June 1991 that I had reopened this case and planned to exhume the remains of Dr. Weiss, reports appeared far and wide, from the New Orleans *Times-Picayune* to the *New York Times* to the *Today* show. Announcements appeared in the *Baltimore Sun, USA Today, San Francisco Chronicle, Chicago Tribune,* and *Washington Post.* This was no picayune or trivial story from out of the misty past. People all over the country were curious about what I might discover. In fact, the *Times-Picayune* ran a cartoon with the amusing caption "Scientists have now confirmed that Zachary Taylor was actually shot by Lee Harvey Oswald after Abe Lincoln poisoned

J.F.K. with a tainted cheeseburger Elvis made for Carl Weiss to give to Huey Long on the grassy knoll in Dallas, according to Oliver Stone . . ."

But I kept the media at bay, for I realized that at that early stage I was too far from an end result even to prognosticate that I would have anything of merit to share with the press or the scientific community when all was said and done.

THE WHEREABOUTS OF THE WEISS .32-CALIBER FABRIQUE nationale haunted me day and night. I even found it difficult to concentrate in my classroom teaching. Where was it or, better, who had it at this fifty-six-year remove from the shoot-out in 1935?

As mentioned previously, the Freedom of Information Act had been a productive source in my research into the provenance of the gun. The picture of the Weiss pistol that Superintendent Guerre had sent to the FBI bore a legible serial number that could not do anything but help my search.

But first things first. I drew up a list of those who had been involved in the Weiss investigation, from the desk sergeant in Baton Rouge to the state police superintendent. It took the tireless and imaginative sleuthing of a New Orleans private investigator, colleague and friend Gary Eldredge, whom I knew from the AAFS, and his staff to narrow the possibilities of who might still possess the weapon. eBay had not yet captured and captivated the field of historical artifactual research.

I had looked to Zinman's book, *The Day Huey Long Was Shot*, which was first published in 1963, in which the author reported that General Guerre had informed him in 1961 that the Weiss pistol "had remained behind at State Police Headquarters after he left" his office there as top man. Yet it seemed to me that the files and gun had been lost around the time that General Guerre had been ousted in 1940 by a new, anti-Long administration. I wasn't about to be misdirected by Guerre's apparent deflection of knowledge and responsibility.

Gary Eldredge was hard at work in squirreling out the gun's secret hideaway when one day, quite unannounced and unexpected, this spry old geezer came puffing up the four flights of stairs to my office at the law school. He must have been eighty-five years old, if a day. He introduced himself as a stringer in Washington, D.C., for the New Orleans *Times-Picayune*. Parking himself on a chair in the midst of my entropy of books and papers, he informed me that his name was Edgar Poe. Edgar Poe! I wondered if his middle initial was A, but I didn't ask. He was there on the business of my investigations into Senator Long's death and he knew what I was up to. Quite to the point without wasted breath, he asked, "Have you found the gun yet?"

"Found the gun?" I said. "No."

"Oh," he said, "I know where it is."

I urged him to tell me, but he countered with "What are you doing to find it?"

I told him that I knew from my earlier cases that law-enforcement people often consider such items to be trophies. With Sacco and Vanzetti, the state firearms expert laid claim to the firearms evidence without telling anyone. So in this case, I suspected the police in the first instance. Among them I was targeting people who were closest to the state police investigation, and Superintendent Louis Guerre, who had supervised the Huey Long investigation single-handedly. Following Guerre, I would suspect each person associated with him or with the case, with Murphy Roden, his right-hand man being second in line.

Mr. Poe listened to me, nodding, and said, "You're on the right track." But he would not tell me who had the pistol, even though, cagey old coot, I was certain that he knew. (I learned only later that he was bound by a journalist's oath of confidentiality.)

But just to hear that I was proceeding in the right direction was a boost, giving me added enthusiasm for the chase.

At the very least, if I could locate the pistol, I could examine it to see if it had malfunctioned, as one witness had reported. We could also try to determine if a bullet from that gun grazed the wristwatch of Murphy Roden, as some have said, leaving striations from the passing bullet locked onto the softer metal of the watch.

It wasn't long before success was mine—thanks to Charlotte Glass of Gary's staff, who at my suggestion had checked the New Orleans court probate record for the probate of General Guerre's will, a paper record of the .32-caliber was found. It was listed by its serial number in the inventory of Guerre's estate as an item of personal property belonging at his death to him and valued at $25. What colossal gall, what unredeemed chutzpah to inventory as one's own that which belongs to another, especially in such a mysterious and historic homicide.

But of General Guerre's three daughters, who among them was the legatee who took possession of the Weiss weapon and who had it today? Performing his usual and unseen magic with public records and wowing investigative wizardry, Gary identified Mabel Binnings, a daughter of General Guerre, as our woman. Her bank safety-deposit box matched exactly that previously rented by her father. Could the gun possibly be there?

I suspected that General Guerre might also have squirreled away the missing state police's investigative files on the death of Long in his private warren as well. They may have been listed in the estate inventory under the tantalizingly uninformative title "Envelope of miscellaneous papers and documents." Those items were assigned "no value" in the inventory, whereas we knew that if they were in fact the police files from the Long/Weiss confrontation they would have inestimable value in the search for the truth about the deaths of Long and Weiss.

Bill Donovan, a homicide forensic investigator with the Louisiana coroner's office and the son-in-law of General Guerre's estate lawyer, who was in 1991 a senior retired Federal District Court judge in New Orleans, went with Gary Eldredge out to Ms. Binnings's home. As they reported to me, they knocked on her front door when a woman asked, "Who's there?"

They identified themselves and told her that they knew she had possession of the Weiss gun. She told them to go away. Uncertain how best to proceed next, they left and telephoned me, bringing this new development to my attention.

This turn of events spoke to me of the inevitability of future litigation to retrieve the gun from Ms. Binnings's firm grasp on it. But action had to be taken promptly, since the cat was out of the bag, giving her the incentive to make haste to dispose of the vital items she might yet have.

I related all of this to Dr. Carl Weiss and suggested a lawyer for him through a connection I had via my son's wife in Opelousas. Leslie R. Schiff, of Sandoz, Schiff, Keating & Watson in Opelousas, Louisiana, agreed to represent Carl Weiss, Jr., by filing a suit on his behalf on September 10, 1991, in Orleans Parish Civil District Court to recover the weapon and to sequester it until his day in court arrived.

The fifty-sixth anniversary of Carl Weiss's death had been that previous Sunday, and due to an intervening holiday, the suit could not be filed until the Tuesday following the anniversary, which nevertheless landed exactly on the anniversary of Huey Long's death.

I believed that all was proceeding as planned toward the recovery of the pistol when Sheriff Paul Valtreau of Orleans Parish called to relate that he had "this court order to go sequester these items."

"Yes?" I said. "How can I help?"

"I've never done this before," he admitted. "This is a civil action. I only do these things in criminal actions. I don't know what to do."

With that as an entrée, I informed him that I thought the gun was in Mabel Binnings's safety-deposit box but that she might have relevant investigative files of the state police on their Long–Weiss investigations. "While you are demanding the Weiss pistol," I suggested, "tell her you want everything that she has that is in any way connected with the death of Huey Long."

Even though reluctant to expand the sequestration order on his own hook, the sheriff agreed to give it a try.

Sheriff Valtreau served Ms. Binnings with Weiss's formal claim to the gun's ownership, asking her to relinquish it to his possession during the pendency of the Weiss litigation. She turned him away and said that the gun was not in her home. He asked if the gun was in a safety-deposit box, and she reportedly responded that it was none of his business.

Two days later, the state police intervened in the petition to recover these items, claiming them as the property of the state. Colonel Marlin Flores, deputy secretary of the Department of Public Safety and Corrections, forwarded a letter indicating that Superintendent Guerre's conversion of state's evidence to his personal use, absent a court order, had been improper. It was legally the property of the state, so it was asserted. A state trooper delivered this letter to Ms. Binnings, giving her until the following Wednesday to respond. He reported that she had graciously accepted it.

The state police then filed a writ of intervention in the action of Carl Weiss, Jr., and an injunction to prevent the destruction or transfer of any and all documents, evidence, or property associated with the Huey P. Long assassination. Suddenly, the state police's fifty-six-year-old quiescent investigation came alive with renewed force and determination. They claimed to have opened the closed files on Long's death, regardless of the desuetude that had set in after so many years.

On September 19, Secretary of State Fox McKeithen joined the legal battle to recover the investigative files concerning the assassination. His office also filed a petition of intervention in civil court to acquire the documents and bring them under protection of the Louisiana State Archives. Notwithstanding my having first located the Weiss gun, I was being deprived of access to it for scientific testing and evaluation. I didn't expect gratitude. My demands were much simpler. I wanted the gun.

Moreover, I felt myself to be a stakeholder with an impartiality the state police lacked in view of their settled view of Weiss's guilt. My wish to take control of the gun did not receive an official acquiescence save for Gary Eldredge's presence when the contents of the safety-deposit box were first spilled into public view.

Once these respective suits were filed, the press moved in again, and articles were published far and wide, in national papers such as *USA Today* and papers local to Louisiana. The press responded to this unfolding drama like Hollywood gossips to the breakup of a celebrity marriage. Everyone wanted to know what would happen next.

Through her lawyer, Ms. Binnings finally turned over to the sheriff fourteen files of papers, the gun, and other items. Under court order, he gave it all to the state police. As they worked to place each page of the documents into a protective Mylar sleeve, no one else was allowed to examine them.

I considered that decision to be legal chaos in action. I wanted to take possession of the gun so that we could ascertain that this was the Weiss .32 and could examine and test-fire it. I did learn that the serial numbers from the FBI's photograph and the gun from the safety-deposit box were the same, so it was clear that the Weiss gun had been found. That, at least, was a victory. Separate from the gun, inside a ladies' hairnet, the sheriff had also found a .32 spent bullet and the gun's magazine with six cartridges in it.

For me, that spent bullet was the single most important item of all. It might be the wedge to pry the door open on the entire tale. That bullet reportedly bore minute traces of what appeared to be blood and two impact sites indicating its having struck a hard object twice, possibly in the course of a ricochet. It plainly appeared to be associated with either the death of Senator Long or at least the firing spree in the capitol corridor on September 8, 1935.

But for the time being, I had to turn my attention elsewhere to another potential source of knowledge: the exhumation of Carl Austin Weiss, M.D.

LOUISIANA POLITICIANS DENOUNCED MY PLANNED EXHUMATION of Dr. Weiss, misunderstanding my aims, which were entirely scientific, and skeptics wondered whether I would next similarly violate the grave of the great Kingfish himself. In fact, some historians believed that Long's exhumation would make more sense, but the multi-ton memorial atop Long's grave fronting the state capitol would have required enormous effort and expense for what I considered to be little payoff—if any. Besides that, his entire family opposed such a procedure. The situation was precisely that which should disable any thought of exhuming the Arlington Cemetery–interred remains of President John F. Kennedy. Despite the impassioned criticisms, I had no intention of exhuming Huey Long or forgoing an exhumation of Dr. Weiss.

The elemental question in any exhumation is the condition of the remains, which, in turn, is often dependent on the type of coffin and any vault that might have been used. I had learned from Merle Welsh that the coffin had been manufactured across the bay from New Orleans at the Orleans Casket Company. I wrote to the manufacturer, who sent me a picture of what the cedar coffin would have looked like before going into the ground. I also learned what I could about the nine-hundred-pound, twelve-gauge, "burglar-proof" Clarkote steel vault in which the coffin was situated.

Dr. Weiss's parents had purchased an expensive vault for him, while they themselves were buried in plain wooden coffins next to their son, without the protective shield of a vault. His monument was also much larger than the simpler memorials provided for the family. This was firm evidence for me that they had believed in his innocence. In fact, Dr. Weiss's funeral had been attended by more people than had ever turned out for anyone in the state who was not a Louisiana mogul.

Earlier in my investigations, I had visited the Roselawn Cemetery in Baton Rouge to probe the ground to determine if the vault was still intact. Mary Manhein, an anthropologist from Louisiana State University, who was on my team, had accompanied me. When the cemetery superintendent got down with his probe and banged on the vault, he said he could tell from the type of noise he heard that there was air in it, not water. But there was also rust on the top, which proved to be less than two feet from the surface.

These early preparations had been satisfactory in boosting my prospects and in gauging the distance between Dr. Weiss's grave and those of his parents. But just in case the side walls of the Weiss grave pit collapsed during the exhumation, I was ready to shore up the sides to protect against a most undesirable disturbance of his parents' graves. Fortunately, the water table was not that of New Orleans, with Baton Rouge being exactly at sea level.

The exhumation of Carl Austin Weiss, M.D., was to serve a number of scientific purposes. It would enable my firearms examiner, Lucien Haag, to evaluate the bullets and fragments that were to be gathered from the remains by the project's forensic pathologist, Dr. Irvin M. Sopher from West Virginia. The autopsy itself was intended to investigate and assess whether some latent physiological or pathological cause might have precipitated or been a contributory factor in the actions of Dr. Weiss on September 8, 1935.

Our team's anthropologist, Dr. Douglas H. Ubelaker, Curator at the Museum of Natural History of the Smithsonian Institution, stood ready to address the anthropological facets of the exhumation as well as to identify the remains as those of Dr. Weiss, and to examine his skeletal remains for signs of bullet strikes or bony anomalies that might have vital significance. He had consulted with me in 1989 on the remains of the victims in the Alfred Packer case and was prepared to join the team on a future project in Massachusetts should I manage to secure permission to exhume Lizzie Borden's murdered father and stepmother in Fall River.

Doug Ubelaker, a taciturn, hard-driving, no-nonsense forensic scientist, had expressed but one concern in signing on to the Weiss exhumation. He was not inclined to wade into the grave pit to retrieve the bones if they were mired in the muck of a watery grave. There are two kinds of exhumations—the skinny or the stinky. He was more inclined to skinnies with bare bones than to stinkies with odoriferous flesh.

Samples of any still-intact tissues were to be collected and submitted to the team's forensic toxicologist, Dr. Alphonse Poklis of Virginia, for analysis for toxins, poisons, and other exogenously induced substances of importance.

As much as we could, we would attempt to determine the site of entry and exit of all bullets, as well as compare those sites against what we knew from the oral descriptions given by the bodyguards and others concerning the shooting of Dr. Weiss. Other experts stood ready as consultants, and Mary Manhein assisted us in obtaining the Lafayette Parish Forensic Laboratory for autopsy uses. Among my support staff, which would include such positions as documentary photographer and project coordinator, were Julie Kempton, David Johnston, James Kendrick, Janet Densmore, and Bill Donovan.

The overall objective was a simple and forthright one: to determine whether Dr. Weiss's remains gave evidence supporting the sworn testimony of Long's Cossacks at the coroner's inquest, or whether his remains cast a pall of doubt over their reports and, consequently, over the culpability of Dr. Weiss in the killing.

Early on the gray and overcast morning of October 20, 1991, with the consents of the Weiss family and the permission of the Roselawn Cemetery satisfying Louisiana's statutory legal requirements for an exhumation, and in the presence of Carl and Tom Ed, I prepared to exhume the cypress coffin containing the remains of Dr. Carl Weiss from its resting place.

A front-end loader removed twenty inches of soil from above the vault and from around its ends, until the pit was forty-two inches wide and sixteen feet long. We dug around the sides to facilitate inspection of the vault's condition, which displayed corrosion but nothing worse. Straps were placed around each end of the vault and attached to the front-end loader. We then drew a black plastic tarp around the site to block it from public view, and at approximately 10:00 A.M. we lifted the vault and casket out of the grave.

It should have crossed my mind, but it didn't, that after over fifty-six years in the ground there was bound to be rust. I soon learned that the vault's lid was rusted shut and the latches would not open to release it. I had invited Captain Lonnie Jones or his representative from the Louisiana State Police to attend as observer. One of their lab supervisors came to rescue me from my embarrassment—or so he thought in good faith. He was armed with a sledgehammer. He slammed it against the vault with all the determination and oomph Oliver Cromwell's men had in knocking the heads off the religious statues in the Catholic churches of Ireland. I heard the wooden coffin slats inside collapsing, the glue binding them together having long since dried and decayed away.

In a panic I visualized the wooden coffin lid falling onto Dr. Weiss's remains, causing incalculable artifactual skeletal damage. Putting my panic into action, I yelled for him to stop, stop, stop. A simple pry bar was summoned, and that was all that was needed to release the vault's latches.

Around an hour after we first lifted the vault, we finally had broken the locks and removed its cover. We saw at once that the vault's inside metal dome had exfoliated and dropped corrosive pieces onto the coffin. Between that and the grave's humidity, the nails had rusted away and the coffin had collapsed, although the wooden slats were intact. To the team's disappointment, we saw that the remains were mostly skeletal, with a few shreds of adherent soft tissue evident. There would not be much of a tangible nature available for our pathologist, Dr. Sopher, for an autopsy. However, the remains were fragile and in danger of sustaining yet more damage if we left them in the coffin for transport, so we worked on removing the surrounding coffin parts without disturbing the remains.

Around noon, we placed the remains into a wooden crate especially prepared for this purpose by the farsighted Bill Donovan and transported them fifty-five miles to the Lafayette Parish Forensic Laboratory. There we recorded more details of the coffin and engaged Don LeCocq to take full-body radiographs from every possible angle.

The X-rays of the skull revealed a large-caliber bullet lying just outside the skull. It may have fallen out in transport, or been jarred loose when the sledgehammer almost sent the vault to kingdom come. Whoever had prepared Dr. Weiss for burial (Merle Welsh?) should have seen it. Luke Haag went over the remains with a mini–metal detector that located dozens of metallic fragments, many not seen as radiopacities on the X-rays.

He also detected two additional intact bullets, one in the pelvis and one in the head. They were .38-caliber, consistent with the pistol that Murphy Roden had used. We took X-rays of the trauma areas, which showed that the bullets were concentrated in the chest area, with one gaping defect in the left cheekbone. Further anthropological analysis remained to be done by Doug Ubelaker at his lab to determine bullet trajectory paths as well as exit and entrance locations of the projectiles that had felled Dr. Weiss.

Even though there were only fragments of mummified flesh on the bones, we did find sufficient brain tissue left inside the skull for a toxicological analysis, which Dr. Alphonse Poklis, our forensic toxicologist, was set to perform.

Mary Manhein worked sedulously at the sink, washing and cleaning the final shreds of tissue off the bones to prepare them for the trip to Washington, D.C.

Doug Ubelaker observed that the defleshed bones would be too wet to transport without the danger of mold developing on them. The risk of any mold's obscuring sites of trauma or otherwise compromising his analyses was too great to chance.

A further complication came when a reliable rumor arose that the state police, always a force to be reckoned with, were about to impound the remains to prevent our leaving the state with them. Their ludicrous continuing criminal investigation gambit had already worked to their advantage in enabling them to gain exclusive access to the contents of the Binnings safety-deposit box. In view of the very real likelihood of the state police's strong-arming us, it was decided that our departure from Louisiana had to be moved up to the next day.

In preparing for our departure, Doug suggested that I lay the bones out in my hotel room's adjacent second bed overnight and turn them every hour or so to dry them. In other words, I was to be sleepless in Lafayette.

I told him that in studying for my exams years before, I had used a trick to keep myself from falling into a dead and imperturbable sleep. I would sleep with a golf ball in my hand, held out over the side of the bed, and when I had relaxed enough to drop it, the sound of the ball hitting the floor would wake me. I would then be ready for another round of study.

"Sounds like a good idea," he remarked, but not very convincingly.

So together we laid the damp bones out on the bed, making no effort to place them in anatomically correct order, and I put a heavy weight in my hand. Throughout the night, I would wake with a start as the weight dropped and, in a woozy daze, turn the bones.

By morning, the bones were deemed to be sufficiently dry to crate for travel. Upon joining the team members downstairs for breakfast, I was asked the usual question: How had I slept?

"I can't say I slept in peace," I replied, "but Dr. Weiss, he slept in pieces," I quipped.

That pun, in my sleep-deprived condition, came back to haunt me in a Baltimore Circuit Court years afterward. I was then testifying as an expert in opposition to the proposed exhumation of John Wilkes Booth when upon my cross-examination the opposing attorney sought to portray me as a frivolous jokester and not as a solid, even solemn, scientist. My quip, which had been quoted in a Louisiana newspaper, was dredged up and cited as signaling my nonscientific frolicsomeness and, therefore, my lack of credibility.

It was a good tactic by a young, bright, and gung-ho attorney, but I survived the challenge to my scientific credentials. Booth was not exhumed and I went on to pun another day. There is a lesson somewhere in this episode. Like Huck Finn, who prevailed over the Widow Douglas's unremitting attempts to civilize him, I have so far prevailed over the ardent wishes of those who want me to fit the mold of a stuffy, humorless forensic scientist. I have found that exhumations are not fitted to the insensitivity of the emotionless, lest the nights surely will be sleepless.

The bones were carefully and individually wrapped for transport and placed into two cardboard boxes, the long bones in one and the torso, skull, and pelvis in another. Once again it came over me like a seventh wave (always the largest, if you know where to start the count) that the whole kit and caboodle of a person's skeletal anatomy can be packed comfortably into just two such medium-sized boxes.

En route to the New Orleans airport, we discussed our options if the state police were determined at the airport to separate us from Dr. Weiss's bones. It was agreed that if trouble brewed and we were given the chance, we would go by land in a rental car, not by air.

The crucial deciding moment would be when our carry-on boxes of bones were X-rayed. At that pre-9/11 time, there were no security or other regulations restricting or prohibiting our taking human bones aboard an airplane. I cannot speak for current Homeland Security proscriptions.

However, if the bones bore flesh, even in limited amounts, they would in 1991 be considered biohazardous materials requiring a conveyance by a funeral director to the point of departure and a pickup by another funeral director at the terminus. Further, forget about taking biohazardous material of this kind on board as carry-on luggage. Knowing of this travel restriction, I was steeped in praise of Mary Manhein's labors in cleaning the bones of the flesh.

As we waited our turn, not a lengthy wait in those days, at the X-ray machine, I decided to send the box with the long bones through first in a trial-and-error test. The other box, with more scientifically relevant bones, was held in reserve just in case the first one did not clear the security hurdle.

All went well for a moment, but as I bent to pick up the cardboard carton the technician asked me what it contained. When I admitted to its being packaged human bones, she gruffly instructed me to send it through a second time. Dutifully but anxiously, I did so while hanging by my fingernails on the result. The box cleared again and the technician motioned to me to collect it.

Even though I had run the gauntlet of security, I couldn't resist asking why she had called for a second viewing of the contents of the box. "Them's human bones," she intoned, her reserve breaking into a smile. "Never seen human bones before. Just wanted to see 'em again."

Then, safely on the plane and a mood of easy-street relaxation settling in on me, I hoisted the box of bones I was carrying into the storage space above my seat and with a sense of relief sat in my assigned seat. But my angst was not yet ended. A man with a rather oversized carry-on bag stopped in the aisle next to me, lifted his bag with his two hands, and was about to stuff it into the same compartment where the box of bones had been placed. I nearly jumped out of my seat, concerned for the safety of Dr. Weiss's bones.

"Don't do that!" I shouted.

He halted in midact and asked, "Why not? What's in the box?"

"Decaying human remains," I said quietly but emphatically.

With a suddenly ashen face, he took his oversized bag and moved down the aisle to another overhead bin, never to converse with me again.

NOW THAT THE PHYSICAL EVIDENCE HAD BEEN FARMED OUT TO the team's specialists, the waiting began for the test results. Dr. Poklis was first to report. He was unable to evaluate the tissues for the presence of a brain tumor, but he was able to perform a drug screen. His conclusion was that no identifiable drugs were present in the tissues and that any identifiable intoxicants that may have caused Dr. Weiss to act irrationally that night would likely have broken down by this time.

It was now Doug Ubelaker's turn to see what tales the bones might tell. He inventoried the bones and verified the age, sex, and race. In addition, he performed a computer-assisted photographic superimposition comparison of the skull with two different photographs of Carl Weiss. These firmly established that the bones were those of Dr. Weiss. Further confirmation by DNA profiling would not have added measurably to what we already knew, even if DNA profiling was then advanced enough to come to our aid.

Ubelaker then set about to map the gunshot trauma to the bones and bullet trajectory paths. He found no evidence of trauma to Weiss's hands from possible impact with Long's face, although he indicated that this did not necessarily disprove reports that Weiss had hit Long with his fist. He found that discernible blackish stains were rampant on the teeth. He identified the stains as mercury, probably from amalgam tooth restorations from which the mercury had leached out after death. No, we were glad to say that we did not have an Alice in Wonderland's Mad Hatter whose mental aberrations and behavioral peccadilloes stemmed from frequent exposure to mercury in working with the leather for hats.

Many of the bones were in pieces, apparently shattered by bullets. Ube-laker glued these bone fragments together, revealing bullet strikes that had not been readily apparent. He examined each of the points of impact in detail to determine under a dissecting microscope the direction and angle of travel of the bullets.

Eleven ribs showed eighteen separate damage sites. Utilizing a hanging skeleton mannequin that he used as a teaching tool, he glued the tip of a drinking straw to each impact site angled in the direction of the bullet's entry.

A bullet hole under Weiss's left eye indicated that he had been shot from a low angle up into the brain case, where the bullet remained until the exhumation and analysis. It was a hollow-point .38-caliber bullet whose hollow point had not mushroomed. Something had decelerated it before it entered Weiss's skull. Otherwise, its muzzle velocity should have sufficed, Luke said, for it to have exited. Since it did not, he suggested that it may have struck something else before entering Weiss's skull.

Luke found a tuft of white linen fibers impacted in the hollow nose of the bullet. Dr. Weiss was seen in the marbled-hall death photo to have been wearing a white cotton suit. So it was a reasonable surmise that the bullet might have torn through the suit. But at what point? The carpal bones of his left wrist gave clear evidence, seen also in his death photo, of a through-and-through bullet wound at that location. Could it be that the wound was inflicted by the bodyguards as Weiss had his arm raised in a self-protective, defensive gesture? That seemed to me to be the most probable hypothesis.

Weiss also had a wound under his chin that Ubelaker thought may have occurred while his head was tilted backward. No apparent exit was noted, but this finding was consistent with Roden's later report—which contradicted his inquest testimony—that he had shot Weiss in the head from under his chin and that he saw the flesh open up.

This second bullet strike to the head could account for the .38-caliber hollow-point bullet found by us in the Lafayette lab lying to the side of Weiss's skull. It, like the bullet we withdrew from Weiss's intracranial cavity, had similar linen fibers embedded in its unexpanded nose.

It, too, must have hit something, decelerating it before it entered Weiss's head. Could the bullet strike to Weiss's right elbow have been the cause of the deceleration? If that scenario was in fact the case, Dr. Weiss had been laid low by two bullets to his head while his arms were raised in a reflexive but futile effort to save himself from his attackers. But the bodyguards, under oath at the inquest, never admitted to having shot him while he was in a defensive posture. Would they have been expected to own up to such surefire evidence of their homicidal intent?

Doug's finding two sites of bullet entry to Dr. Weiss's skull and no exit wounds brought into sharp focus Coroner Bird's report of only one entrance wound and one exit wound. Was the coroner just mistaken, or was something more nefarious afoot? In any event, our findings were indisputable and irreconcilable with Coroner Bird's conclusions.

Ubelaker, thorough and conscientious as ever, computed the angle of the many shots and identified a minimum of twenty distinct bullet strikes with the possibility of up to twenty-four, if those that struck the hands and arms did not also strike other bones. Some additional bullets may have pierced only the soft tissue without traversing bone. To that we could not speak.

The trajectories he mapped suggested that twelve (50 percent) of the bullets had been fired at Weiss from behind, seven from the front, three from the right side, and two from the left. Since Dr. Weiss had been shot so many times from at least four different angles, we had support for the possibility that Long was caught in the intense cross fire and accidentally shot.

None of the bodyguards testified that Weiss had been in a defensive position at any time or that they had shot him from behind (in fact, they disclaimed doing so), but our evidence strongly supported the opposite. At the very least, then, these findings threw into severe and serious doubt the credibility of Long's skull-crushers at the coroner's inquest. Should their self-serving declarations then, or should the sure and straightforward and unbiased voice of Dr. Weiss's bones now, be controlling?

IN A HEARING ON OCTOBER 25, WHILE THE BONES WERE STILL being unpacked and analyzed in D.C., the New Orleans court granted the Louisiana State Police permission to test-fire the Weiss .32 using one or more of the six Rem-UMCs (Remington–Union Metallic Cartridges) in

Ms. Binnings's inventory. Patrick Lane, a state police firearms expert, conducted the test in the presence of Luke Haag, as the representative of my team, and Captain Lonnie Jones, representing the state police. The results turned out to be unforeseen and startling, like physiologist Andrew Nalbandov's discovery in 1940 that he could conduct long-term hormone research after the pituitary gland of chickens had been surgically removed because the janitor cleaning his laboratory left the lights on overnight. Sometimes scientific findings are just the luck of the draw.

The .32-caliber spent bullet that had been in the safety-deposit box was determined by Lane's comparison microscopic examination to his test-fired bullets not to have been fired by the Weiss .32 automatic.

The reason for the exclusion of the Weiss .32-caliber semiautomatic as having fired the enigmatic bullet from the Binnings's collection was a surprising, uncommon, and incontrovertible exclusion. It was an exclusion not based on differences in the minute striations (the distinctive barrel markings) of the bullets being compared, but in the difference in one indisputable class characteristic: the land widths. Both bullets were .32-caliber, and both possessed six lands and grooves with a left twist. So far they were comparable on their class characteristics. But the irreconcilable incompatibility arose from the differences in the land widths on the test bullet and the evidence (Binnings) bullet.

A disparity in land widths could be readily explained away if the evidence (Binnings) bullet had land widths that were narrower than the land widths of the test-fired bullet. It is always arguable that the land widths might change with use so that a later-fired bullet would explicably have a land width wider than a previously fired bullet. Under those circumstances, the firearms expert would have to resolve the dispute of the incompatibility of the land widths based on the further evidence of the individual or accidental markings (striations) on each bullet distinctive to the barrel of each gun that fired it.

But there was no need to go further than the disparity in the land width characteristics in this case. It was akin to saying that a .32-caliber bullet cannot be fired from a weapon chambered to fire only .45-caliber ammunition.

The tipping point here was the recognition that the land widths on the Binnings .32-caliber bullet were *larger,* not *smaller,* than those on the Weiss test-fired .32-caliber bullet. Whereas over multiple firings land widths may expand, they cannot and do not contract. The result: The Weiss gun must be excluded as the weapon that fired the Binnings .32-caliber bullet that is presumably the fatal bullet, the injury from which caused Huey Long's death.

But Patrick Lane and Captain Jones did not see it that way. When Lane looked through the eyepiece of his comparison microscope, you could feel his palpable consternation, so Luke Haag reported to me. In Jones and Lane's seeking to escape the dismay of what the microscope revealed, it was asserted that the bullet in question had nothing to do with this case. But how could such a turnabout on their part be squared with their prior belief in Weiss's guilt?

That statement just did not measure up, because the Binnings bullet had all the trappings of involvement in *l'affaire* Huey Long. It bore discernible traces of calcium carbonate, a major constituent of marble (the halls were marble). It had a detectable impact site in the nose from striking a hard object like a marble wall. It had brushing or burnishing of the jacket's tin plating that typified its passage through a soft object (like bodily tissue and clothing), causing it to be scored and be decelerated from its original muzzle velocity of about 850 feet per second to only about 200 feet per second.

These were not only guesses or surmises plucked from the air. Luke Haag himself test-fired a similar .32 weapon with vintage .32-caliber Rem-UMC ammunition against witness panels of similar construction as the marble walls. The results were at all times the same—the Binnings bullet was not unrelated to the Long shooting. It was not a vagrant or a migratory bullet from some other event, such as an unconnected test firing of the Weiss gun in 1935 by the police.

The evidence was staggeringly compelling that the Binnings bullet had been fired at and into and through some person on September 8, 1935, in the capitol corridor in East Baton Rouge, Louisiana. If the person struck by it was not Huey Long, then who was struck, there being no one other than Long who had sustained comparable injuries? Roden claimed to have a cut on the wrist where he wore his watch.

If this was the bullet that killed the Kingfish, then Weiss could not have fired the shot with his .32-caliber Fabrique Nationale. Someone else who had a .32 weapon had to have been the shooter, but the identity of that person and the whereabouts of that gun were a mystery, and probably will remain so. Even though Long's Cossacks might have fired at Dr. Weiss with much-larger-caliber weapons, still .32-caliber weapons were a featured and common pocket-sized backup weapon for the police at that time. The failure to take into account the possibility of the police's carrying and using an offbeat backup weapon is what caused the erroneous first report of SLA (Symbionese Liberation Army) leader Donald DeFreeze's (his *nom de guerre* was General Field Marshal Cinque) suicide in the 1974 Los Angeles shootout from which Patty Hearst gained such notoriety.

Rhetorically speaking, if the Binnings bullet had not killed Long, then of what significance was it? Why had it been saved along with the gun, magazine, and six cartridges by State Police Superintendent Louis Guerre— which, I repeat, are the only items of firearms evidence extant from the 1935 melee in Baton Rouge?

Since we did not have the bullet (or bullets) that had been removed— if, in truth, any had been—from the Kingfish, we could not compare it to the bullet that had just been fired from Dr. Weiss's gun.

Unfortunately, the weapons wielded by the bodyguards have disappeared into the never-never land of lost or mislaid physical evidence. Thus, we could make no comparison tests with any of those.

In short, more tests and investigations to complete the circle of guilt would be desirable if and when any new firearms evidence surfaces, but for now and for the long haul, Dr. Weiss cannot be connected to the assassination of Huey Long through the firearms evidence.

DOUG UBELAKER AND I PRESENTED THE ANTHROPOLOGICAL FIND-ings privately to the Weiss family, giving them reinforced hope that Carl Austin Weiss, M.D., would be fully and finally exonerated.

I was scheduled to present our findings in February at the 1992 meeting in New Orleans of the American Academy of Forensic Science when I learned that a local judge had issued a gag order on all who had witnessed the firearms examinations in the presence of Lane and Jones. Apparently, the judge wanted to one-up for himself the newsworthiness of my announcement.

I went ahead with my presentation notwithstanding, since I was not present at the firearms test firing and analyses nor was I said to be named in the gag order, a copy of which was never served on me. My lawyer background told me I was on safe legal grounds in going forward with my report.

As I spoke to a full-house audience, I noticed the presence of two police officers in mufti standing in the rear of the lecture hall. You always know by his bearing, his stance, and his authoritarian carriage when you are in the presence of a police officer.

After my lecture, the state police firearms expert, Patrick Lane, rushed to the platform and warned me in a quavering voice that he would see to it that I was arrested for being in contempt of the gag order. Having been a Korean Conflict veteran and having survived other scrapes with death, I did not readily fold in fear. But the press, having overheard Lane, wanted an arrest, possibly more than Lane.

For the next day and through the night, the press followed me everywhere, even to the point of camping out in the hallway outside my hotel room. They smelled blood, my blood, and they wanted to be in on the kill. But their waiting and tracking were all for naught. I was not arrested; nor was I bothered, bewitched, or bewildered by law-enforcement officials.

Once again, with the luck of my Irish forebears, I had escaped unscathed. Later exhumations, such as that of CIA agent Frank Olson, would pose even greater hazards and potential challenges from the thin blue line.

I later learned that the records found in Ms. Binnings's cache included not the hoped-for police documents but a detailed investigation of an alleged conspiracy behind the assassination, which shed no real light on the shooting. However, the recovered files did contain photographs of the suit, shirt, and undershirt that Long had worn that fateful night, which revealed one entrance wound in the front and one exit wound out the back (where the Binnings bullet had exited?). Sooty residue around the entrance hole indicated that the shooter had been close, possibly within an inch or two. So the one-bullet theory appeared to be correct, but I still did not know from whence came the fatal bullet, except to say with apodictic certainty not from Dr. Weiss's .32-caliber gun.

In June 1992, a local judge ruled on the application of Leslie Schiff, lawyer for Carl Weiss, Jr., that Mabel Guerre Binnings had no legal right to the weapon or the records of the state's investigation of the Long assassination. They remained, for the time being, in the care of the state police archives. Eventually, the Weiss gun was returned to its rightful owner, Dr. Carl Weiss, Jr., the son of the deceased. He, in turn, donated it to the State Historical Society. I have heard that it is on display in the basement of the capitol in Baton Rouge, bearing a note that credits me for having found it.

As a result of these investigations, both under and above the ground, the guilt of Dr. Weiss for the assassination of Senator Long became palpably improbable. The bodyguards were proved to have been either grossly inaccurate or deliberately misleading in their descriptions of Senator Long's death and of their shooting of Dr. Weiss in the corridor of the capitol. In addition, the firearms testing established conclusively that the .32 spent bullet, the only item of firearms evidence cached away by State Police Superintendent Louis Guerre after the shoot-out in 1935, was not fired by Dr. Weiss's Fabrique Nationale.

The unanswered question, therefore, the *terra incognita* of this investigation, is resolved into a single question: If Dr. Weiss did not kill Senator Huey Long, then who did? Obviously, it would have to be one of his Cossacks during the shooting from all sides in the corridor. But which of the Cossacks was the most likely candidate? Were the killer's actions merely reckless, or was the motivation less excusable?

Solving one tantalizing mystery, the supposed guilt of Dr. Weiss, opened the door to another mystery. But who has the will or the way to follow that trail? Certainly not the Louisiana State Police, whose glare has been fixed (fixated?) on Dr. Weiss since 1935 as the only conceivable culprit.

Sometimes not even the illumination of forensic science can dim the glare or muffle the sound of police precognition as to a person's guilt or innocence.

3. FRANK OLSON

The Man Who Fell Thirteen Stories

We grow accustomed to the Dark—
When light is put away . . .

The Bravest—grope a little—
And sometimes hit a Tree
Directly in the Forehead—
But as they learn to see—

Either the Darkness alters
Or something in the sight . . .

EMILY DICKINSON

He was a conflicted man living in conflicted times who died leaving conflicting leads.

When a forty-three-year-old man falls thirteen stories from a hotel room window in midtown New York City to his death, it may seem to be a daunting task to determine the true circumstances of his death, especially after his being buried for more than forty years. Yet after exhuming and examining his remains, a specific injury that went undetected at the time of his death turned out to be an important lead in reevaluating the manner of the man's death.

Not only had the elite government scientist whose death was in question *not* died in the manner officially described, his body did not bear the injuries stated in his autopsy report. With the discovery of these new and troubling facts, many questions arose, such as whether this "suicidal leap" (the deceased being referred to in the parlance of officialdom as a jumper) had been related to his employment resignation, which was offered only a few days before the incident. Had he found out something he was not supposed to know?

These questions remained for my scientific team to try to answer, and once the victim's family asked for my aid, told me what they knew, and then warned me, "Now, of course, you know you're taking on the CIA," I began walking with my back to the wall, any wall would do for cover.

What seemed to be at first glance a simple and straightforward set of events leading to a death turned out to be much more complicated than had been anticipated, including the occurrence of yet more mysterious deaths.

THE VICTIM UNDER SCRUTINY WAS FRANK OLSON, PH.D., WHO had died sometime after midnight on November 28, 1953, on the Seventh Avenue sidewalk outside New York's Hotel Statler. Robert Lashbrook, Ph.D., who was with Olson on that fateful night, recounted his version of the events leading up to the tragic moment many times: Lashbrook and Olson were together in Room 1018A on the tenth floor—actually the thirteenth when the first three unnumbered floors are counted—when Dr. Olson suddenly went out the room's only window. In his initial version of the events, Lashbrook insisted that he had been asleep at the time, and that he awoke to the sound of crashing glass. Only then did he realize that Dr. Olson had catapulted through the room's closed window, apparently bent on suicide.

A New York City assistant medical examiner, Dr. Dominick Di Maio, confirmed Lashbrook's recital, stating in his report of his external (not internal) examination of Olson's body that there were many compound fractures of Olson's extremities and multiple lacerations on his face and neck as well as his lower extremities. This report was filed, was uncontested, and as such, became the official record. Yet there were perturbing features surrounding Olson's death that only gradually came to light over a span of many years during which the Olson family tried to unlock what they were convinced was a more terrible truth about his death.

Eric Olson, Dr. Olson's elder surviving son, was nine years old at the time of his father's unexpected death. On the morning after the incident, he learned from his father's government boss, Colonel Vincent Ruwet, that his father had died in a work-related accident that had occurred in a hotel. He had either "fallen or jumped," but young Eric could not understand the full importance of what that meant. Over the years, as he came to think about the claim that his father's death resulted from a "fatal nervous breakdown," he discovered that his father had been an employee of the CIA engaged in sinister and clandestine research activities of a biochemical nature.

The Olson family was in 1953 living on the verge of Frederick, Maryland, while Frank Olson was employed as a biochemist at nearby Fort Detrick, the Army's premier bacteriological-warfare research installation. Since 1943, he had been part of a team of scientists who were immersed in a top-secret program aimed toward developing lethal biological and chemical weapons for America's defense during the Cold War, a subject considered to be a matter of utmost secrecy for the protection of national security.

In 1949, Frank Olson had helped to set up the Special Operations Division (SOD) at Fort Detrick, where written records were forbidden and only a trusted few were allowed to know about the more sensitive projects. Olson was tasked to develop new and secret biological means for effective interrogation and warfare. He soon became the acting head of this division. Among its projects, according to what Eric's research taught him, were the development of assassination materials, collaboration with former Nazi scientists, LSD mind-control research, and the use of biological weapons during the Korean War. The ominous nature of such mind-control research was exposed to public view in the Hollywood movie *The Manchurian Candidate,* starring Frank Sinatra and Laurence Harvey.

Eric remembers that his mother was uncomfortable about the work her husband was doing. He is of a mind that his father had expressed distress over experiments he was conducting and the possible use of the results. While a great deal about his activities remained unknown to his family, during the weekend preceding his death Frank Olson apparently had been uncharacteristically distraught, he being a person of considerable cheerfulness and bonhomie. He had come home early from a meeting at a mountain retreat and had said something to his wife, Alice, about "a very bad mistake" he had made while presenting a paper at the meeting. He seemed deeply and singularly anxious about it, but he would not reveal the nature of his mistake. According to Mrs. Olson, he felt certain that his career was in jeopardy and he had decided to resign.

Yet when he returned from work the next day, he told Alice that his colleagues had reassured him. Consequently, he withdrew his resignation letter. However, he said it was recommended that he obtain treatment for some newly emergent behavioral problems he was having. The plan of action included his seeing a psychiatrist. Otherwise, he said, she might not be safe with him under the same roof. Alice Olson was stunned. Nothing in his demeanor had indicated that he was dangerous. But he did not explain what he meant any further. It was only two days before Thanksgiving 1953, but notwithstanding the importance of his family holiday, he insisted he needed to leave for treatment. He believed he would return in time for Thanksgiving dinner.

The next day, an official SOD car came to convey Frank Olson to Washington, D.C. Alice joined her husband during the thirty-plus-mile trip, then saw him enplane for New York. He was accompanied by his boss, Vincent Ruwet, and a stranger who was introduced to her as Dr. Robert Lashbrook.

In New York City, at Fifty-eighth Street, Olson reportedly had several sessions with a medical doctor that lasted most of the day. He and Dr. Lashbrook had Thanksgiving dinner in New York, but Olson telephoned his wife on Friday and told her he expected to be home that Saturday. He was contemplating entering a psychiatric hospital in Washington for treatment. But his plans were to be thwarted. That night he died in New York City. He was forty-three, and left behind a wife and three young children, two sons, Eric and Nils, and a daughter, Lisa.

His death being ruled a suicide, his body was sent home to Maryland for a burial. His casket was at all times kept closed due, so the family was informed, to the massive injuries he sustained in his fatal fall. Frank Olson was buried in a solemn ceremony on December 1, 1953, in Linden Hills Cemetery in Frederick, Maryland, where a stone monument to his memory was put in place.

For twenty-odd years, the Olson family remained in a state of perplexity about these events. In her grief, Alice Olson took to drinking to excess, ultimately becoming an alcoholic.

His father's sudden death haunted Eric, who was now the man of the house. He felt certain that his father had not deliberately jumped out the window, but he was at a loss as to how to resolve his suspicions.

Then, on June 11, 1975, a front-page article in the *Washington Post* engaged his excited attention. Now, after the passage of so many years since his father's death, the political scene in Washington, D.C., had changed. No longer were people afraid of Soviet biological warfare. In fact, many people had expressed distrust of the government's conduct of the Cold War and were rooting out past secrets, some of which were shameful and unethical. This *Post* newspaper report, revealing the results of the Rockefeller Commission's findings about illegal CIA activities, noted that a civilian employee of the Department of the Army had jumped from the tenth floor of a New York hotel after he was surreptitiously given the hallucinogenic drug LSD. "This individual," the article stated, "had not been made aware he had been given LSD until twenty minutes after it had been administered." It had been part of a larger CIA-orchestrated project that had involved the administration of psychoactive drugs to numerous unsuspecting Americans.

The victim's name was not disclosed, but the description of the incident was too similar to the circumstances of Frank Olson's death for Eric to ignore. He showed the article to the rest of his family. He proceeded with direct inquiries to Vincent Ruwet, who had been his father's supervisor at Fort Detrick. Indeed, Ruwet had visited with Alice Olson often after her husband's death, purporting to sense and share her grief.

Ruwet reluctantly admitted that the unnamed man in the *Post*'s article was in fact Frank Olson. He had been a guinea pig in a CIA-sponsored testing of LSD at a retreat at Deep Creek Lodge in rural Maryland. Afterward, when he seemed to be suffering from serious side effects, the CIA had decided to seek treatment for him, taking him to New York for that purpose. Before it was completed, however, he had jumped to his death, so Ruwet maintained.

In righteous outrage, the Olson family invited the media to a press conference in July 1975 to announce their intention of suing the government for its complicity in the wrongful death of Frank Olson. It was hoped the media attention would pressure the CIA to make a full disclosure of Olson's tragic death. Alice made the first statement, indicating that she believed her husband must have been suffering from some sort of bad dream to have jumped—if that is what happened—through a closed window. Then Eric recalled that the family had never been told any of the details, that they had been callously deceived, and that they wanted to know why a cover-up had been in place all these years.

Four days later, an invitation to the White House arrived. In a meeting that lasted less than twenty minutes, President Gerald Ford offered a complete and uncompromising apology and urged the government to grant the family three quarters of a million dollars as a monetary settlement. The President also ordered CIA Director William Colby to cooperate with them.

When the family, in due course, met with Colby, who later reported that this had been among the most difficult assignments he had ever had, he offered them 150 pages of redacted documents that he claimed amounted to the entire file relevant to their concerns. It was the CIA's investigation into Frank Olson's death. He believed it would answer any questions they might have. He seemed unaware of the importance of what he was giving them, for the material soon raised more questions than it answered. It became clear to them that there was something darker in this tragic incident than a failed LSD experiment.

PORING OVER THE DOCUMENTS, IN WHICH MANY LINES AND names were blacked out, the Olsons believed that significant information about their father's death was still missing. There was no explanation in these pages, for example, about why the CIA agents had bundled Frank Olson off to a New York City hotel rather than a hospital. If he had been in such a self-destructive frame of mind, why was his supposed escort asleep when he exited the room through the window? And why, when he went out the window, had Lashbrook failed to summon help or even notify the hotel's staff? The family now came to the painful realization that Frank Olson might have been murdered. But if so, what was the motive for such a nefarious deed?

In their questing for the truth, they discovered that CIA higher-up Sidney Gottlieb, Ph.D., who was officially reprimanded for the handling of the Olson incident, had been involved in CIA assassination plots on national leaders and was an enthusiastic supporter of mind control. Gottlieb's experiments occurred at a time in our country's history when no act performed with a view toward thwarting communism was considered out-of-bounds. The standoff between the two world powers meant that survival was seen as the end that justified all means, including the development of biological weapons and mental manipulation. Intelligence reports had the Soviets on the fast track and the Americans playing catch-up. The CIA's activities were disguised in projects bearing innocuous names that concealed their actual purpose.

Passages in the documents they had received pertaining to a CIA operation called "Project Bluebird" and renamed ARTICHOKE were deleted, but Eric Olson plumbed the depths from other sources. ARTICHOKE had involved extreme methods of interrogation and an attempt to develop a way to produce complete amnesia in questioned subjects or in agents who had seen too much and could no longer be trusted. It was also disclosed that Olson was not just a "civilian scientist" with the Army but a full-fledged CIA employee. He had been an agent in a powerful position and had possessed detailed covert information.

The laboratory experiment in which Frank Olson had unwittingly participated had been part of a "truth drug" program, supervised by Sidney Gottlieb and George Hunter White, that involved getting people to disclose all under the influence of a drug clandestinely administered. If the CIA found the right drug, they could use it to extract secrets from enemy agents as well as learn how to protect their own agents against such disclosures. They had started with the active ingredient in marijuana and moved on to more dangerous drugs, like LSD. Once the program directors decided that informed subjects could not give authentic results, agents had administered LSD in large doses to unsuspecting soldiers at the Edgewood Arsenal and to unconsenting civilians in hospitals. The CIA managers offered emoluments to universities soliciting their involvement. For the purposes of this research, some people were kept in a hallucinogenic state for days at a time.

At the New York Psychiatric Institute in January 1953, less than a year before Olson died, other experiments had been conducted. Harold Blauer, a tennis professional, went there for depression. He became one of the guinea pigs, but his reaction to the LSD spelled disaster for him. After a bad reaction, he succumbed to a coma and in short order died.

Operation Realism, a top-secret project run by George White, involved giving citizens in bars and restaurants LSD without their knowledge. White even set up massage parlors for Operation Midnight Climax, another CIA program, as a way to lure people into their LSD experiments.

Frank Olson was described in these heavily edited pages as one of the CIA's guinea pigs, making him a victim. But being so involved in its operation, he would hardly have been an easy target. Nevertheless, the report described how on November 19, 1953, Gottlieb or someone acting at his behest had slipped the drug into Olson's glass of Cointreau at Deep Creek Lodge. After twenty minutes, Olson was said to have developed hallucinations, after which he was told that he had been a subject in their experimenting. By morning, he was still in an agitated frame of mind.

After his return home in a dispirited and depressed state, he went to work and told Ruwet of his intention to resign. Ruwet indicated to investigators that Olson had appeared to be "all mixed up." He and Lashbrook then took Olson to New York. Instead of being put under the care of a qualified psychiatrist, Olson was taken to Harold Abramson, an allergist with a CIA clearance who was a firm believer in the therapeutic value of LSD for psychiatric patients. At one point, he apparently gave Olson bourbon and the sedative Nembutal, both central nervous system depressants whose adjuvant effect could have killed Olson. The sleep they might have induced in him could have been his last.

By some accounts, Olson might also have met a magician by the name of John Mulholland, who may have tried to use hypnosis on him. Ruwet told investigators that Olson became highly agitated and paranoid while in New York. He spent one night wandering the streets, and at one point he discarded his wallet and his identification papers, asking to be allowed to "disappear." Ruwet said that Olson did not want to go home to face his wife. Yet the next day, he called his wife to assure her that he was better and expected to see her the following day.

Lashbrook reported to the police who were investigating Olson's death that he, Olson, worked for the Defense Department and that Olson had been calm that evening, washing out his socks in the sink before going to bed.

Yet four hours later, Olson fell, so Lashbrook said, to his death.

As part of the police investigation, Lashbrook was taken to the Fourteenth Precinct station house, spending only a brief period there. He told the police that he did not know why Olson had killed himself, except that he did suffer from ulcers. The detectives asked him to empty his pockets but did not keep a record of what they found. However, a Security Office report indicated that he had airline ticket stubs for the trips that he and Olson had taken, and a receipt for $115 dated November 25, 1953, and signed by John Mulholland. Supposedly, this was an advance for travel to Chicago.

Lashbrook also had hotel bills and papers with phone numbers, including those for Vince Ruwet and Dr. Abramson. In addition, he had an address for a house on Bedford Street that was used for Operation Midnight Climax. One sheet of paper had New York City addresses for people identified only by the initials G.W., M.H., and J.M. Lashbrook said that for security reasons, he preferred not to reveal who they were. The detectives apparently did not press the matter.

Since John Mulholland had died in 1970, there was no way for the Olsons to interview him about his possible involvement in this tragedy. However, he had been under contract to prepare a manual, "Some Operational Applications of the Art of Deception," that applied the magician's art to covert activities, such as slipping drugs into drinks. There was no record of the actual manual's having been produced.

Right after Olson's death, the CIA sent five investigators to New York without explaining why they would send so many in the case of an outright, uncontested suicide. An internal memo the following week refers to Olson's "suicide"—in quotation marks—as if the memo's author was aware that it had not been a suicide. And one of the phone numbers that Lashbrook carried that fateful night was for George White, the man in charge of the program, whose alias was Morgan Hall. Lashbrook's immediate boss was Sidney Gottlieb. Although George White operated a CIA safe house in Greenwich Village, only minutes from the Hotel Statler, where Olson and Lashbrook had taken lodging, they apparently did not visit it.

The investigating team recommended disciplinary action against

Lashbrook and Gottlieb, but whereas Lashbrook left the agency, Gottlieb remained in power for the next two decades. (He is said to have dismissed Olson's death during hearings in 1977 as one of the risks of running such experiments.)

The Olson family was disturbed by what they had learned. Although the men who were investigated had claimed that Frank Olson was in a suicidal frame of mind, they had roomed him on a high floor. He had even managed to slip away from them one night, a good indication that they were not really watching him. On Thanksgiving Day, they had found him in a shell-shocked state in the hotel lobby. Dr. Abramson had diagnosed him as psychotic and recommended hospitalization. Yet he remained in the hotel for two more nights.

Piecing these facts together, Eric Olson thought that Lashbrook's account was implausible. He decided to investigate on his own. So in 1984, he went to the Hotel Statler (now the Hotel Pennsylvania) to see the room for himself. It was a basic hotel room with two double beds, small and rather spare in its furnishings. He could not imagine how anyone could have gotten a running start in such a room without awakening the person in the next bed—the man who was posted there to watch Eric's supposedly delusional and suicidal father. The sill was high and there was a radiator right in front of it. The shade had also been pulled down. He wondered if it was in the realm of possibility to break through the window with so little space allowed to gain momentum and so many obstructions at the window.

Experts later told Eric that a man would have to be running more than thirty miles per hour to crash through such a window and that the hotel room was too short for even an accomplished athlete to accomplish it.

Another mystery that seemed to be part of Olson's November death involved the trip that he had taken overseas the prior summer. Again, the family had to piece together different sources of information to understand his travels. He had been to Scandinavia, Germany, and Britain, that was certain. But for what purpose?

In London, Eric learned from a reporter that his father had talked with an expert on brainwashing about something he had witnessed at research installations in Frankfurt. Eric was led to believe that these facilities tested human subjects called "expendables"— who were enemy agents or collaborators and who sometimes died from the experiments. Perhaps Frank Olson had voiced his dismay and disgust over this inhumane behavior and was for that reason considered a potential security risk. Alice recalled that when Frank returned from Europe that summer, he was unusually withdrawn and morose and contemplative.

To Eric, the newly revealed facts painted a grim picture of his father's death having been a CIA-staged suicide. The CIA operatives may have slipped the LSD into his father's drink to get him talking, and once they saw his reaction, decided to get him to New York where the suicide could be faked convincingly.

That was the hypothesis he felt was cementing itself into place.

Yet even before this new angle could be explored more thoroughly, the Olson family was visited by yet another tragedy—one that was indirectly instrumental in bringing me into this steamy John le Carré–like brew. To tell that story, I ask the reader's indulgence as I turn the clock back a few years.

THE OLSON FAMILY AND I WERE LINKED IN A FRIENDLY, SOCIAL relationship through Greg Hayward, a student of mine at The George Washington University's law school. Greg had married Eric Olson's sister, Lisa.

Greg was a rugged and totally engaging outdoorsman who had served as an Airborne Ranger in Vietnam. We were biking companions, and also went rock climbing and rappelling. He owned a farm in Frederick, Maryland, to which I would journey with my family on many occasions. But during that time I knew nothing about the death of Frank Olson or the family's sustained grief over it, a grief that they keep well hidden from me.

Greg was readying himself for a run for Congress. I have no doubt his knowledge and likable personality would have carried the day for him. Lisa was active as a teacher for the deaf. They were a wonderful, much-loved couple with a young son, Jonathan. It was a joy to know them.

One day in 1978, Greg came to my university office, his face glowing in a way that suggested he had the best of news.

"I'm here to tell you," he said, "that the government has paid the Olson family close to a million dollars for their wrongdoing in causing the death of Frank Olson."

Surprised and curious, I invited him to tell me more about it, and so he did, putting me on notice of the circumstances of Frank Olson's death. I congratulated him on the government's fessing up for their wrongdoing and his finding solace in it. The moment was one to be savored by him and the Olson family.

Shortly thereafter, Greg called to invite me to join him in an airplane trip to the Adirondacks, where he had a cottage. I was sorely tempted to join him and his family, since I had once before visited his remote cabin and felt the memory of that good time tugging me to go.

But I was compelled to decline, my university classes demanding my attention. Reluctantly, he accepted my decision. He signed off with one of his favorite expressions: "Remember, Jim," he recited, "danger is no stranger to an Airborne Ranger." As it turned out, that was the last conversation I ever had with Greg Hayward.

The same night, I received another phone call from another former student, Hugh Lewis, stating that the plane carrying Greg to the Adirondack retreat had encountered a sudden, unexpected snowstorm, causing it to crash into a mountainside. Both Greg and Lisa, their two-year-old son Jonathan, and the child that Lisa was carrying had died in the crash, the entire family wiped out.

And if I had not been so stuffy, so wedded to my faculty duties, I would have gone, too. This tragedy stayed with me, as did my questioning the reason for the fate that had saved me. Would there come a time when I would be called to repay my good fortune?

Then, in 1993, Alice Olson died. This was the impetus for Eric to act more aggressively on his unflagging suspicions over his father's death. A knock once again came at my office door from a member of the Olson family. This time it was Eric, who was familiar to me from my cycling excursions to Greg Hayward's farm.

Eric explained to me that his mother, Alice, had been buried in Mount Olivet Cemetery in Frederick, Maryland, while his father, Frank, had been interred in 1953 in Linden Hills Cemetery, also in Frederick. It was his wish, which he articulated with his usual enthusiasm, to remove his father's remains to Mount Olivet Cemetery so that, in death, they could be reunited. But, at the same time, he was quite obviously suffused with another, even more compelling, desire.

Eric represented in a very straightforward way and with a calm and determined mind that he and his brother, Nils, wanted to have their father's remains scientifically scrutinized to see if modern methods of analysis could provide tangible evidence of the underlying cause and manner of his death. "We want to find out if we can learn more than we already know," he emphasized.

Eric's request, even entreaty, was entitled to and received my immediate and careful attention. I cautioned him that I would coordinate and oversee the project only if my exacting criteria for an exhumation were fulfilled. He was entirely agreeable to that arrangement.

From what I could determine from many background sources, there was certainly substantial controversy over whether Frank Olson's death was an accident, a suicide, or, more ominously, a homicide. Further, the fact that no autopsy had been performed on Olson militated in favor of our discovering new and possibly determinative evidence to shed new light on his death. It was also clear that all of the immediate family supported this project.

The first step involved a word-by-word examination of the New York City medical examiner's report, rendered in 1953, on the death of Dr. Olson. With the necessary consent of Eric Olson, I secured a copy, which revealed its author to be Dr. Dominick Di Maio, later to become the New York City medical examiner. His report was most abbreviated, since the death had been "no posted" (reported based only on an external examination of Olson's dead body). The accompanying toxicological report assayed only the presence of methyl and ethyl alcohol in the liver, with negative results, and included no drug scan of any kind. Those were strong indicia that a thorough autopsy could potentially accomplish something more than was previously done—contingent on the condition of Olson's remains. That condition, for good or for ill, was the most uncertain and most significant aspect of this, as it is in any exhumation, especially where the time elapsed since burial is prolonged.

Seeking to flesh out the particulars of Di Maio's written report, I telephoned him in New York City and found him to be cordial and candid. He conceded that no X-rays had been taken at the time of his examining the remains. As he put it, he had been "taken in" by the reports he received that this death was an uncomplicated out-and-out suicide. Indeed, in the 1970s, after the *Washington Post*'s revelations, his ire over learning that he had been misled resulted in his contemplating revisiting the death of Dr. Olson. But other matters had preempted his time.

These discussions with Dr. Di Maio, along with my review of the vast documentary evidence regarding Dr. Olson's death and the congressional investigations into it, convinced me that an exhumation was warranted.

Three main objectives loomed largest in my approach to this exhumation:

1. To give the Olson family confidence that all available scientific and investigative means had been employed to bring the truth about Frank Olson's death under public and scientific scrutiny.

2. To provide a forum for the utilization of new or underutilized scientific technologies and experiences, such as the bioengineering aspects of a fall from a height, an analysis of the causal features in fractures resulting from such a fall, a toxicological analysis of bodily tissues and hair for therapeutic and abused drugs (whether defined as "controlled substances" or not), the use of the computer to animate a reenactment scenario of the event, and to provide an identification of the remains by a computerized skull superimposition.

3. To examine and to ponder whether the tragedy of scientific experimentation with the lives and well-being of unwitting persons that marred and stigmatized this CIA research enterprise might conceivably be replicated in today's society.

But first it was necessary to obtain written and notarized authorizations for the exhumation and analysis of the remains from Eric and Nils Olson, as the two surviving children. Fulfilling any legal requirements in Maryland for an exhumation was next in my line of investigative fire. As I have found in other states, the Maryland statutory provisions governing exhumations are haphazard, spotty, and unclear. For clarity I went to a presumed reliable source—the state's medical examiner's office in Baltimore. After explaining my interest in removing Frank Olson from one cemetery for reburial in another cemetery, beside his wife, I was directed to the local district attorney for Frederick County.

That contact, by phone and mail, was cordial and cooperative in the sense that no court order approving the exhumation was deemed necessary. Of course, it is fair to say that if the full details of the exhumation had been demanded of me, a different attitude and result might have ensued. The autopsy and its sequelae, which were to intervene between the exhumation and the reburial, were not featured in my approaches to the legal authorities in Maryland. Whether the investigation would have been stymied or sidetracked if all had been told I cannot say, but it is probable that Frank Olson's remains would not have been housed aboveground under lock and key in my office and in that of Dr. Jack Levisky at York College, Pennsylvania, over the nearly ten-year span that transpired until Olson's reburial.

With these preliminaries accomplished by October 1993, I proceeded to assemble a team of qualified and eminent specialists in the multiple scientific disciplines that would be put to the task in this investigation.

The total came to fifteen. Dr. "Jack" (James) Frost, a West Virginia deputy chief medical examiner, agreed to perform the autopsy at Hagerstown (Maryland) Community College, arranged through the contacts made by Jeff Kercheval, a criminalist with the Hagerstown police lab who served on the team. Geologist George Stephens, Ph.D., of The George Washington University was in charge of any geological assessments. Yale Caplan, Ph.D., a former president of the American Academy of Forensic Sciences, would perform the toxicological analyses. Michael Calhoun, a radiographer with Shady Grove (Maryland) Adventist Hospital, would do the X-raying. Jean Gardner, Esq., stood by for her legal insights, along with three photographers, including Gerry Richards, a retired chief of the photography section at the FBI and a former student of mine in the MFS curriculum at The George Washington University.

Last to mention, but in the forefront throughout the project, was

Dr. John "Jack" Levisky, a forensic anthropologist and department chairman at York College, Pennsylvania. A scientific consultant and an assortment of support staff were also used. Some of these people would become regulars with me in my future exhumations.

I provided the background on the case to each person, and all of them responded with alacrity to my invitation, even after hearing the many complexities and, possibly, imponderables that lurked on the horizon. Each team member deserves the highest praise and has my most genuine and entirely unreserved gratitude for their selfless, uncompensated labors in our attempts to disentangle this mystery.

All was now in readiness.

I organized the project in my usual bilevel manner. On one level, the belowground one, it was strictly a scientific investigation. On another level, the aboveground one, it partook of investigations into the facts and circumstances of Olson's death. In combination with the scientific findings, the aboveground investigations would contribute to a better understanding of how Dr. Frank R. Olson came to his death. Those aboveground investigations would include securing interviews with people with knowledge of the incident.

My foremost interest was to locate the whereabouts of persons present at the Hotel Statler on the night Frank Olson died. I focused on the Statler's assistant night manager, Armand Pastore, who first found Dr. Olson lying supine and barely alive on the Seventh Avenue west-side sidewalk; the Statler's telephone operator, who overheard Dr. Lashbrook's call from Room 1018A following Olson's exiting the window; the priest from the nearby Roman Catholic Church who gave Dr. Olson last rites as he lay dying on the sidewalk; and the New York City police officer on the beat who responded to the death scene. As the investigations progressed, others would be added to the list of interviewees, including the CIA's mandarin for clandestine research projects, Sidney Gottlieb.

While the aboveground investigations proceeded, the exhumation was given the go-ahead.

The exhumation at Linden Hills Cemetery took place on Thursday, June 2, 1994, commencing at 8:30 A.M., under clear skies and in the presence of the Olson brothers, my team members, representatives of the press, and a smallish group of voyeurs. We removed the metal coffin from the grave without incident, then transported the unopened coffin to the nearby (thirty-plus miles west) Hagerstown Police Department Crime Laboratory for the unveiling of the remains and for Michael Calhoun and his wife, Sue, to perform their radiographic wizardry.

On Friday, we transported the remains to the Biology Department at Hagerstown Junior College, where we would conduct the autopsy, analyze the skeletal features, and obtain specimens for toxicological analysis.

When we opened the coffin, the remains proved to be immaculately well-preserved, albeit mummified, under a tight wrapping of linen acting as a full-body shroud. We were able to look at the body as if the incident had happened only the day before. We did not observe the slightest sign of mold or decay. One expects that—after some forty years in the grave, despite embalming—the coffin's interior dampness from the moisture resulting from the release of body fluids would have produced some mildew and that the body would have begun to putrefy.

Yet there is never any assurance as to the condition of remains over time; the type of coffin used and the preservative qualities of the embalming fluid, keeping decay at bay for sixty years plus or minus, are just two factors to take into account. I have seen corpses in much worse condition with a lesser time period in between burial and exhumation. But such was not the case here, probably because of the lack of the disruption of an autopsy on the remains and the care taken to preserve the remains for interstate transfer from New York City to Frederick, Maryland.

On other occasions, the remains of persons long dead have been exhumed and, surprisingly, even startlingly, seen to be fully fleshed and well-preserved although mummified. The finding and the opening in Paris, France, in 1905 of the leaden coffin of American naval hero John Paul Jones, some 113 years after his death, is one such instance. His preservation is best explained by Ambassador Porter, who reported to the secretary of state that "the body fortunately was found well-preserved, the coffin having been filled with alcohol but which had evaporated." Of course, as with Frank Olson, the linen winding sheet in which both he and John Paul Jones had been wrapped was added protection against postmortem decay.

In the first instance, our attention was drawn to Olson's face and neck.

I saw no lacerations there, nary a one, although the autopsy report had in-dicated the presence of "multiple abrasions and lacerations." While we were examining the remains prior to the autopsy, Eric Olson arrived and insisted on viewing his father's remains. Although it is standard practice to keep the relatives out of the autopsy room, I made an exception in this case after noticing the uncanny resemblance between Eric in life and his father in death.

Seeing Eric standing beside his father's remains gave me a start, for one was a look-alike of the other, in the same way that civil rights leader Med-gar Evers and his son were seen to be duplicates of each other when Medgar Evers was exhumed by Dr. Michael Baden from his grave in Ar-lington National Cemetery.

The Olson family had been advised against an open-casket funeral cemetery on the advice of Dr. Olson's superiors, who had represented that his injuries were too gruesome to behold. Yet on our close inspec-tion, there was no evidence of such injuries. That was a matter of some considerable consternation. How could the autopsy report have stated the existence of multiple lacerations when, in truth, there were none?

Further, the entire anterior (front) portion from head to toe of the flesh of Frank Olson's remains was devoid of lacerations, save those to be expected from the many compound fractures, and from that in the upper thoracic region where a laceration had been sutured up, probably during embalming.

The significance of this absence of lacerations is less a criticism of the first autopsy than a commentary on the way Dr. Olson exited through the window. Certainly, going through an open window would be one feasible explanation, because if it had been closed, it is reasonable to expect that the glass would have cut his skin at some place or other. If a drawn shade separated the glass from the flesh, then perhaps the shade kept the glass from piercing Olson's underwear-clad body, but it is nevertheless inexpli-cable that we found no cuts on the front of the lower extremities from dragging across glass shards on the bottom edge of the window.

The literature reporting on broken-glass-type burglaries reveals that the broken glass causes lacerations most frequently when the body, generally an arm inserted through the broken glass to open a door or window to facilitate the burglar's entry, withdraws.

For the moment, we could only say that it is most probable either that Dr. Olson went to his death through an open window or that he went through a closed window with a shade drawn in front of it. The lack of lacerations gives only an immeasurable edge to the open-window hypothesis.

We turned our attention to the head. It is altogether improbable that Dr. Olson would have gone unresisting to his death if he were conscious of a third person's trying to force him out the window to a certain death. His resistance or, rather, the reaction to quiet his resistance by such third persons might have left its imprint on the skull of Dr. Olson. Our sights were therefore particularly set on that portion of his anatomy.

Yet before we could go further with these examinations, we had to establish the identification of the remains as that of Frank Olson. It may seem superfluous for us to have labored to identify the remains from the grave, marked by a toe tag as Frank Olson, as those of Frank Olson, but such are the intrinsic necessities of a scientific investigation. The grave marker was presumptive evidence of his identity and burial location, as was the toe tag attached to the remains in the coffin.

We knew that Dr. Olson had been a pipe smoker, which could be reflected in his teeth, that he might have taken some falls from horseback riding, and that he'd been discharged from the army for an ulcer. These facts could become helpful for the purpose of solidifying the identification. In exhumations, no stone should be left unturned, for there will probably be no future opportunity to revisit an exhumation once the remains are reburied.

Jeff Kercheval and Sherry Brown, of York College, rolled the mummified palmar surfaces of the finger pads after having infused them with a saline solution to puff up the flesh, and I Mikrosil-cast them (with a Silly Putty type of material), both with splendid results, in the hope that antemortem fingerprints of Frank Olson in his military records might be retrieved for comparison purposes. However, we were not to be blessed by such good fortune, since his fingerprints, if they had ever appeared in his military records, had been destroyed long before.

Jack Frost, our pathologist from West Virginia, performed the autopsy. Jack is an active, wiry man, a contemporary of mine who has more than once tracked the story of an undetermined death to resolve its particular manner. His instincts are superb and his work ethic exacting and conscientious.

In a full autopsy, the external evaluation consists of examining old injuries along with tattoos and scars. Trace evidence, such as hairs and fibers, is collected off the body and from under the fingernails. Even the nails are clipped or retained in full.

In the standard autopsy, the pathologist makes a Y incision, cutting into the body from shoulder to shoulder, with the arms of the Y meeting at the sternum (breastplate) and then going straight down the abdomen to the groin. A saw (often a Stryker saw) is used to cut through the ribs so that the rib cage can be lifted away as one piece from the soft internal organs.

The next step is to take a blood sample from the heart for drug and other subsequent tests, then start taking the organs out, one by one, to examine and to weigh them. If there is fluid in an organ, it gets drained for a sample, and then the stomach and intestines are opened to examine the contents. We were interested in the condition of the organs of Dr. Olson only for the purpose of assaying injuries to them.

Injuries are generally categorized as blunt-force trauma, gunshot, and sharp-force trauma. In all cases, the number of wounds is recorded and each wound is carefully measured and its characteristics described.

A blunt-force injury comes from impact with an object lacking sharp edges, like a gun butt, a hammer, or, in this instance, the Seventh Avenue sidewalk. A medical examiner will try to determine the direction of impact, the type of object that caused it, and how often contact was made. A suspect weapon may or may not be available, but if it is, then the wound patterns may be connected to the instrument causing them, described as a pattern-type injury.

Sometimes lacerations result, which is a tearing injury from impact and has ragged or abraded edges, often with bruising. There may also be abrasions, or friction injuries that remove superficial layers of skin. Contusions are ruptures of small subcutaneous blood vessels. Crushing wounds result from blunt violence to skin that is close to bone, causing these wounds to bleed into the surrounding tissues.

With gunshot wounds, the pathologist looks for distinct patterns that indicate the type of weapon used, where the bullet entered and exited (if it did), and how far from the body the gun was when the shot occurred.

With knife, or "incised," wounds, the medical examiner must draw a distinction between cut and stab or puncture wounds, and among different types of piercing implements, such as an ice pick and a knife.

Some victims are asphyxiated, which results from cutting off oxygen to the brain. Hanging, obstruction of airways with some object, smothering, or strangulation can cause asphyxia, and each has specific manifestations. Carbon monoxide poisoning can also be a cause of death by depleting the oxygen-carrying capacity of the blood.

After examining and describing wounds for the documentary record, the medical examiner takes swabs from all orifices and cuts pieces from the organs to place on slides for further trace evidence, including firearms, toxicological (poisons), and histological (cellular) studies. Samples of hair are taken as well, and if the body has recently died, urine is removed from the bladder for drug testing. The scientific studies following the autopsy can take weeks to months. Consequently, we knew those results would likely be a long time coming.

Finally, we look at the head, first examining the eyes and skin for pinpoint capillary hemorrhages (called petechiae) that may reveal evidence of strangulation. After that, we incise the scalp behind the head and carefully reflect the skin over the face to expose the skull. Using a high-speed oscillating saw, we open the skull and remove the calvarium (the top of the skull above the brow ridges) so we can lift out the brain to examine and weigh it and inspect the intracranial walls of the skull.

All of this, of course, depends on having a well-preserved body and organs, which we did.

As Dr. Frost systematically worked his way over the body, it was seen through the X-rays that Olson's right foot had taken the greatest impact upon colliding with the pavement. There was a small laceration in the bottom of the right heel, causing a fracture of the calcaneus (the heel bone). The right tibia (lower leg bone) was massively fractured, resulting in its being in two linear pieces. On the left side of the body, there was an indication that the bottom of the foot had hit something hard.

During all these close and careful scientific perambulations, my over-riding concern was with the goings-on in Olson's room that might have precipitated his fall from it. If Eric Olson's suspicions of homicide were correct, then it was possible that Frank Olson had been hit with some-thing in the room and then thrown or dropped out the window. The most likely source revealing such a happening was Olson's head.

Over the left eye, underneath unbroken skin, was a fist-sized hematoma embedded in the subgaleal sheath. This subgaleal hematoma necessarily re-sulted from the hemorrhage of a blood vessel over the left eye. The flesh in that area of the scalp was intact, having experienced, as best we could tell, neither a laceration nor an incised wound, nor was it patently abraded. No fracture of the skull or any other hemorrhage could be directly related to the impact causing this hematoma. In a way, it seemed to stand apart and alone. It had not been noted in the first autopsy report.

Yet Jack Frost thought that if Olson had been hit with a blunt-force object in the room, he would expect to find some indication on the skull's surface. He thought what he found was more consistent with the man's head hitting a broad, flat, firm obstruction—possibly the window frame or the window's plate glass itself. However, the size of the injury suggested to me that if Olson hit his head against a wall or window frame first, he would have been unconscious when he took another run at the glass. And if he butted his head against the glass on his way out, breaking it in the doing, then he'd necessarily had his head up and was looking where he was going with his arms at his sides.

I didn't find it entirely unlikely that someone would want to see what was about to happen to him, but I did think it unlikely that such a suicide-minded person would have his arms at his sides, as Jack Frost's opinion suggested, with his head taking the brunt of the collision with the win-dow. It was improbable that Olson's exit was as Jack Frost hypothesized, especially with the shade drawn and the window closed. Not even Super-man has been pictured soaring off into the blue in that fashion.

To get a closer analysis of the bones and flesh, we transported the remains an hour north to Dr. Jack Levisky's anthropology lab at York College in York, Pennsylvania, where more intensive analysis could be conducted under his supervision.

Chair of the Behavioral Sciences Department there, Dr. Levisky is a quiet and unassuming scientist with a lively interest in historical cemeteries in the York County area. Located close to Gettysburg, Pennsylvania, where a fierce four-day battle was waged in 1863 that turned the tide of the American Civil War, he is fully apprised of local efforts in historical preservation.

In Dr. Levisky's laboratory at York College, the remains of Dr. Olson were kept in a vault under lock and key while Jack Levisky and his assistant, Sherry Brown, worked to macerate (deflesh) the bones. With the bones defleshed it would be possible to gain closer insights into the fractures and their patterns and interconnectedness than even the X-rays provided us. As it happened, some hairline fractures, undetected on the radiographs, were discovered only after inspection of the skeletonized remains. The objectives of the project thus required that the bones be defleshed.

Dr. Levisky's calculations revealed three separate fracture sites: One was in the upper body, particularly the ribs and shoulders, where Olson must have collided with a construction barrier at street level; another was in the lower legs as they impacted the sidewalk, with the forces of velocity having traveled upward to cause a "book" fracture in the pelvis (opening it like the pages of a book), dissipating the upward thrust so as to protect the vertebral column from any dislodgement; and third, a horizontally aligned fracture to the right temporal and parietal bones of Dr. Olson's skull, occasioned in all probability when his head hit the sidewalk after his feet first struck the pavement and he just toppled over. At the point of impact, the skull was split from right to left across the top of the skull, allowing it to be manipulated as if it were hinged.

From our anthropological appraisal, we concluded that in his descent, Dr. Olson's upper body had first struck an obstruction, and we learned later that a wooden barrier had been in place that night, located along the building wall for sandblasting purposes. It was likely he had struck that, spinning his body about so that he landed feet first on the sidewalk in an almost upright posture. Then he fell like a rag doll to his side, coming to a full stop lying supine on the sidewalk.

Yet nothing had been seen thus far to account for the large hematoma to Olson's head. It remained for us to assess how and where it had come into existence, because we were convinced it would be signally instrumental in resolving the dispute over whether his death was an accident, a suicide, or worse yet, a homicide.

To assist in our evaluation, we excised the skin around the site of the parietal laceration over the ear and studied it microscopically. Viewing it under a stereomicroscope left no doubt that this laceration was a unit and not the sum of more than one blow at the same site inflicted at different times or places.

Then we looked for trace elements in the excised tissue that might resolve some of the mystery as to how and where it came into being. A scanning electron microscope, coupled with an energy-dispersive X-ray, disclosed the presence of a number of elements, one of which was silicon in a rather high concentration. The element silicon, second only to oxygen in its abundance in the earth's crust, is a principal component of glass. The supposition surfaced that the laceration showed evidence of a fragment of glass of micron size (one thousandth of a millimeter).

Yet even supposing that this trace of silicon demonstrated the presence of glass, the question remained whether it was glass from the window of Room 1018A or from some other source. And if it was determined to be glass from the window of Room 1018A, there remained the question of whether it came to adhere in this laceration when Dr. Olson's head struck the window or when it hit the window glass already lying on the sidewalk of Seventh Avenue. Of course, it bears mentioning that it could have been random broken glass on the sidewalk having no connection to Dr. Olson's death.

Since the laceration and its subjacent skull fracture plainly seemed to be linked to a single impact, and since that impact would have to have been a blow well beyond the capacity of a window or its housing to inflict, I believed that the glass in that laceration must have originated from a particle of glass picked up at the street level.

But whether that glass was the window glass of Room 1018A or some foreign piece of glass of infinitesimally minute size that had been lying about on the Seventh Avenue sidewalk of the Hotel Statler are equally possible, and that issue is a conundrum not answerable without exhaustive and costly scientific testing. And to what avail?

Our next port of call for identification purposes lay in the teeth. Not having antemortem dental X-rays of Frank Olson, how were we to make an acceptable identification? Photographs of Olson showing his teeth through a smiling face were assembled and forwarded to the team's forensic odontologist, Dr. John McDowell, at the University of Colorado Medical School in Denver. At the same time, the actual skull from our exhumation was forwarded to Dr. McDowell. His report of his dental comparisons was thorough and constituted additional strong support for the remains being those of Frank Olson, even though it lacked the element of surprise that his dental expertise revealed from the exhumed teeth of Jesse Woodson James, as explained in the next chapter.

While we had no reasonable doubt about the identity of these remains, we decided to go one long stride forward with new technology still on the scientific cusp to attempt a computerized superimposition of the skull to known photographs of Frank Olson. Calling on the combined services of Dr. Vernon Spitzer, also of the University of Colorado in Denver, and Michael Sellbert of Engineering Animation, Inc., we were able to use computer wizardry to see how the antemortem photographs of Olson matched the skull from the grave. These telling results left no further doubt that the remains we had autopsied were those of Dr. Frank Olson.

A computer animation of the most likely scenarios for Frank Olson's fall was to follow after additional on-scene measurements based on the recollections of those present at the scene on the occasion of Olson's death fall.

LATER IN JUNE, I LOCATED AND VISITED WITH ARMAND PASTORE, the Hotel Statler's assistant night manager in 1953, who had been summoned to Olson's side. His recollection of the location of Dr. Olson's body on the sidewalk fronting the hotel was quite fresh and keen, giving us a base from which to synthesize a computer animation of the event. It also provided a backdrop for our reconstructing the scene at the street level of the hotel. He said that when he found Olson, his eyes were open while he was making a supreme effort to speak, but his words were only an incomprehensible gurgle. As he lay on his back, one of his legs was twisted at a terrible angle. Before a priest and an ambulance arrived, Olson died in Pastore's arms.

Pastore then went across the street to look at the upper floors from which Olson had doubtlessly fallen. He espied a window shade sticking out through one of the windows in the upper floors—a window that was broken. Having identified the room as 1018A, he proceeded to check the hotel records and found that two people were staying in that room: Frank Olson and Robert Lashbrook. Upon the arrival of the police, Pastore escorted them to the tenth floor. As Pastore placed his passkey into the door, the police drew their guns as a precautionary measure.

Upon the door being opened by Pastore, the room, its window broken, appeared to be empty, aswim in cold air. Lashbrook, the other occupant of the room, was found to be sitting in the bathroom on the toilet with his head in his hands. Lashbrook, giving the impression of being befuddled, said he had been sleeping when he awoke to the sound of crashing glass.

Pastore told me that Lashbrook's statements and behavior disturbed him and led him to doubt the truth of Lashbrook's recitals. Why hadn't he called downstairs to the desk to find out about his friend's condition? Why had he placed a most curious telephone call to another CIA operative stating only that "he" was "gone"? And why was he just sitting there many removes from any positive action?

Another peculiar aspect of the incident was the condition of the bed in which Olson supposedly had been asleep. The bedclothes were pulled back in a way that was consistent with someone ripping him out of bed forcibly.

His suspicions aroused, Pastore spoke to the hotel's phone operator, asking if any calls had come from Room 1018A. She recalled one call. The man in the room had called a number out on Long Island while she had listened in. When an unidentified person answered, the caller said, "Well, he's gone." At the other end, the reply was brief and unemotional. "That's too bad," he said. Then they both had hung up, with nothing more being said. This report did nothing to quiet Pastore's suspicions.

A later report from the CIA, which Eric Olson had brought to my attention, disclosed that the operator had overheard a call from Lashbrook to Dr. Harold Abramson's clinic on Long Island. Abramson was the man charged with overseeing Olson's treatment in New York City.

Other suspicious circumstances emerged. Olson was rushed into the autopsy room, but then they decided not to do a full autopsy. We also learned that the priest who came to the site for the purpose of giving Dr. Olson last rites was quietly moved aside. So there were many aspects of this case that were very suspicious, and they would become even more so.

While we interviewed Pastore, he showed us the approximate location of a wooden barrier that had been in place at the time Olson fell. It was on the sidewalk in front of the Penn Bar, adjacent to the place where he had found Dr. Olson's body. I gained confirmation for some of Pastore's remembrances from another former Hotel Statler employee, employed, upon my meeting with him, at the relocated Penn Bar. He was on duty at the hotel the night of Olson's death. It was also his recollection that there was a barrier in front of the Penn Bar during steam cleaning of the building's street-level stone facing.

From what Pastore could recall, I determined that Dr. Olson had landed at a particular location on the sidewalk, and Pastore told us that when he looked up, the drapes and shades from 1018A were flapping in the breeze outside the window. This firsthand information fixed the location of Dr. Olson's fall and enabled us to reverse his travel, for reenactment, from that point back to the room. We used that information to assist in a reconstruction.

We had a triad of possibilities to consider: that someone in the room had hit Olson and pushed or dropped him out; that he had hit something on the way out or down; and that he had received his injury on the sidewalk. We would use the injury and other information to decide which it was.

Two matters of a bioengineering nature that drew our attention deserve separate mention here. The first relates to the speed at which Dr. Olson exited the window of Room 1018A. The other concerns the physical difficulties to be encountered in falling through a closed window or its housing.

The distance from the sidewalk to the windowsill of Room 1018A was about 173 feet, as determined by triangulation made by geologist George Stephens and me from actual measurements we made on-site. Using Pastore's information and our measurements, we had the engineers from Engineering Animation, Inc., create a computer simulation of the fall, figuring in the height in feet to Room 1018A with the location of the beds in the room and the floor measurements in Room 1018A to determine both the horizontal and vertical velocity of Olson's exiting the room and falling to the street.

It was calculated that for Dr. Olson to have struck the sidewalk-level wooden barrier, causing him to land where Pastore said, his exit velocity would have to have been no more than 1.5 miles per hour, because a greater speed would have propelled his body beyond striking range of the barrier. Consequently, he had fallen at about half the speed of a normal walker's pace, hardly running as if his death depended on it. Yet this estimate of horizontal velocity did not shed light on whether Olson went through an open or closed window, or one with or without a drawn shade. Those factors would have an impact on his speed in exiting the room. Would one and a half miles per hour be sufficient speed for him to exit through a closed window? That remained a lingering and unresolved puzzle.

And still the possibility could not be discounted that someone in the room had inflicted a blow to Dr. Olson in the process of stunning him into submission, preparatory to ejecting him from the window, especially if the window was open at the time and broken thereafter to coincide with Lashbrook's contemporaneous statements.

The next step in our reconstruction efforts was to assay the size of the room, the location of the beds in the room, the height and dimensions of the window, and any obstructions that might have been in front of it. It was within our engineering objectives to seek an answer to the question of the velocity needed to exit the room, whether the window was open or closed at the time. In addition, we learned that the nature and thickness of the window glass, although altered at the time of our on-scene investigations, were identical to that of a window in an upstairs unit at the hotel.

More important, the window in 1018A was unchanged in its dimensions from 1953, and even the radiator that had fronted the window then had been replaced with a similar-size heating unit in the same position. I also learned upon a visit to the window shade's manufacturer in Alexandria, Virginia, that the shade, if drawn, would have impeded Dr. Olson's exit, making it implausible that he could have exited with it drawn at the time. All of these factors affected our computerized reconstruction of the incident and augmented the complexities and the imponderables of bringing the truth of Dr. Olson's death to light.

We know that the horizontal divider on this double-hung window was five feet ten inches from the floor of Room 1018A. We accepted as a fact that Dr. Olson himself was five feet ten inches tall, necessitating his bending over to some degree to clear the divider. Our measurements also revealed that the lower window ledge was thirty-one inches from the floor, requiring a person bent on throwing himself headfirst through the window to elevate himself and be airborne to clear the ledge. The combination of these two gymnastic feats leaves it highly problematic that striking the window glass or the window's housing could realistically be said to have been the cause of the hematoma situated just over and traveling to the orbit of the left eye.

To gain a measure of certainty on this matter, we would have to experiment with a number of five-foot-ten-inch bodies (or simulations of them in the form of mannequins) hurtling through a window of the dimensions and location of the one in Room 1018A. Ruefully, no such experiment could be designed with any approach to verisimilitude. No would-be A students in any of my classes in law or forensic science were game to be guinea pigs in such an experiment. The one component that we cannot factor into such an experiment and without which such an experiment would be fatally flawed is the agony of disorientation that one must surmise overpowered Dr. Olson if he went through the window by his own unreasoned choice.

We gave serious consideration to whether the hematoma to Dr. Olson's skull had occurred at the street level when his body, determined to have been traveling in excess of sixty miles per hour, struck one or more hard and unresisting objects. I discussed this matter with our team's bio-engineer, and he left me with the distinct understanding that if this part of Dr. Olson's frontal bone had come into direct contact with an object at ground level at any speed above eleven miles per hour, the frontal bone would have suffered a fracture and the skin would have been lacerated. None of that having occurred, that possibility seemed to be excluded. In other words, had the hematoma been caused by hitting the sidewalk, there would have been much more damage to the skull and adjacent flesh than we found.

Regarding this hematoma, however, we must address another matter. Hemorrhages of the head do occur from trauma inflicted at a site removed from the site of the hemorrhage. Intracranial contrecoup hemorrhaging is a classic example, as is the hemorrhaging of the eye from gunshot wounds to the head, resulting in what is known in the parlance of forensic pathologists as "raccoon's eye." A contrecoup injury within the intracranial vault occurs when the skull has sustained trauma to one side—say, in a fall—but the interior hemorrhaging is massed on the opposite side of the intracranial vault. The blow to the head creates a wave effect on the brain tissue, causing it to flow to the side of the skull opposite the injury site, where hemorrhaging occurs when the moving brain tissue finds no outlet and crashes against the inner table of the skull. A contrecoup hemorrhage is the consequence.

The specific question, then, was whether the subgaleal hemorrhage over the left eye could reasonably be attributable to the impact to the right parietal bone, resulting in the hinge fracture of the skull.

I noted that the hematoma over Dr. Olson's left eye did not bear even the remotest resemblance to a hemorrhage resulting from a contrecoup cephalic insult. Likewise, Olson's hemorrhage over the frontal bone had no association with the firearms-injury phenomenon that produces a raccoon's eye. That is not to say that this hematoma could not have resulted from the force fracturing the right parietal bone, only that more than speculation must be demonstrated to reach that conclusion.

I REALIZED FULL WELL THAT THE MAJOR UNRESOLVED MYSTERY continued to be the large hemorrhage above the left eye. All of the team members agreed that it most likely had occurred as the result of some incident in the room, not from hitting the sidewalk or in Olson's descent. The question was, Did it happen in the room prior to his going out the window or did it happen at the window when he went out? Given what we knew about Lashbrook's immediate reaction that night, I thought a straightforward, uncomplicated interpretation was likely to be the correct one. Investigations of suspicious events are justifiably reliant on an Occam's razor approach in reaching their conclusions. When two or more possibilities exist, sometimes even when equally plausible, it is incumbent to adopt that rationale which is the simplest, the most direct, and the least convoluted, so Occam's razor teaches.

Among the few matters on which we can only speculate is the likelihood that the traditional view (at least since 1976) of Dr. Olson's exiting the window—glass, shade, and all—is scientifically and realistically plausible. This is a matter that is just not scientifically testable in view of our not knowing and not being able to reconstruct, if we did know, the state of mind of Dr. Olson as he hurtled headfirst through the window on his own (or that induced by LSD) misbegotten choice. For myself, I am solidly skeptical of anyone in Dr. Olson's allegedly distraught state of mind, or even someone of sound mind and memory, clearing a thirty-one-inch-high window opening obscured by a drawn shade, all in the darkness of a hotel room at night, without having his line of travel so obstructed as to cause the venture to misfire. On this issue, until this query is put to rest by adequate scientific testing, I am a die-hard skeptic.

It came time to probe other avenues of possible information on Dr. Olson's death. Testing for the presence of drugs in the bodily tissues and the hair of Dr. Olson was a first order of our scientific sequelae following the autopsy. But what drugs or their metabolites should we seek to discover?

Self-evidently, LSD (lysergic acid diethylamide) was at the very top of our list, but other drugs were on it as well. With all the outrage over the CIA's LSD experimentation, little attention has been paid to the government's testing of other drugs, many of which also have hallucinogenic effects—some, like benztropine, or BZ, ever more potent in even smaller doses than LSD.

After a thorough perusal of the various available Freedom of Information documents of the CIA's drug experiments, I narrowed the field of candidates for our drug testing to tetrahydrocannabinol (the active ingredient in marijuana), mescaline, morning glory, radioactive LSD, LSD, and benztropine. Highlighting these drugs or drug sources, I personally delivered a variety of well-preserved, carefully logged, and continuously refrigerated bodily tissues harvested by Dr. Frost to Dr. Yale Caplan, one of our forensic toxicologists who had established a private, commercial toxicology lab in Baltimore. Hairs from a number of bodily locations on Dr. Olson, some fitted with a root structure, were also submitted to Dr. Caplan.

His written report came back to me in two parts, with the LSD testing considered separately from other therapeutic or abused substances. The testing of bodily tissues for substances other than LSD was negative. The LSD testing of these tissues through radioimmunoassays (RIA) was deemed inconclusive. I telephoned Dr. Bruce Goldberger, Dr. Caplan's associate and the man from whose testing these results were derived. From him I learned that the RIA tests had all given positive results for the presence of LSD in the tissues. However, he added, these results were so unusually uniform as to be considered unreliable. Something about the RIA kits was deemed to be plainly out of whack.

Dr. Goldberger, an internationally recognized toxicologist with the University of Florida, recommended further confirmatory testing at the one location that possessed the knowledge and the technique sufficient to give a firm determination. However, this added testing would cost a minimum of $2,000. I knew that if the Olson sons had only a vendetta against the CIA, the report of inconclusive results that we had in hand would add a dollop of scientific certainty to their anger, leaving them disinclined to go further with drug testing, but much to their credit they agreed to fund the new testing. This proved their stated interest in getting to the root cause of their father's death, whether the findings implicated the CIA or not in something more nefarious than an accidental death.

On my instruction, Dr. Goldberger forwarded select body-tissue samples to Dr. Rodger Foltz at Northwest Toxicology in Salt Lake City, Utah. Dr. Foltz is an eminent toxicologist well versed in the use of gas chromatography coupled with tandem mass spectrometry to assay the presence of LSD in bodily tissues. His report confirmed the absence of any traces of LSD in the submitted samples. Happily, the extraction process had hit no snags. The detection limits set by Dr. Foltz for his analysis were so low that it can be said with no hint of uncertainty that those tissues were, at the time of testing, devoid of any evidence of LSD.

That, however, is most definitely not an affirmation of the lack of LSD in those tissues at the time of Dr. Olson's death in 1953 or prior to it. The dosage of LSD the CIA concedes it administered to Dr. Olson was so minuscule—70 micrograms, according to CIA project director Dr. Sidney Gottlieb—the half-life of LSD is so short (only a few hours at the outside), and the lability of LSD in embalmed tissues over time is so unknown that not finding LSD is absolutely nonprobative on the issue of whether Dr. Olson ingested the substance in 1953.

AT EVERY TURN, THIS INVESTIGATION BROUGHT NEW ISSUES TO light to burden and perplex us. One of these perturbing matters was the question of Dr. Olson's mental stability prior to the Deep Creek experiment on November 19, 1953. According to an undated statement made by Lieutenant Colonel Vincent Ruwet, Olson's supervisor at then Camp Detrick, "During the period prior to the experiment my opinion of his state of mind was that I noticed nothing which would lead me to believe that he was of unsound mind." This assessment was said by Colonel Ruwet to be founded on professional opinions and CIA contacts with Dr. Olson and his family, which he described as "intimate" over a two-and-a-half-year period prior to Dr. Olson's death.

Similarly, Dr. John Schwab, another of Dr. Olson's supervisors at Camp Detrick, gave a statement from the perspective of one who knew Dr. Olson both over a long period of time and at home and in the office—that "his general state of mind and outlook on life was always that of extreme optimism. Never was there any indication of pessimism."

On the contrary, Lyman Kirkpatrick would report on December 1, 1953, that in a conversation with Dr. Willis Gibbons, a higher-up in the CIA, Gibbons indicated that "Olson has a history of mental disturbances." His source for this knowledge was not revealed.

In piecing out the puzzle of the reasons for Dr. Olson's death, the state of his mental health prior to and at the time of his surreptitiously being given LSD is of prime importance. Mental health professionals recognize that in certain persons suffering from mental distress the ingestion of LSD can have consequences that are reflected in altered personality traits and behavior. Yet the weight of the evidence from friends, family, and professional colleagues of Dr. Olson is that he was outgoing, even extroverted, a family man devoted to his children and in all respects a well-balanced individual prior to the night of November 19, 1953, at Deep Creek. But Dr. Gibbons and the CIA, in questionable apologies following Olson's death, would have us believe otherwise.

A well-documented effect of LSD use in some persons is the occurrence of flashbacks days and even months after the LSD has been taken. However, the voluminous literature on the hallucinogenic effects of LSD, compiled from a time before that of Dr. Timothy Leary, the LSD guru of his time, to the present, demonstrates well-nigh conclusively that during flashbacks LSD users do not commit violent acts such as throwing themselves through a closed window. Even a government study commissioned on account of Dr. Olson's death and published by the Government Printing Office makes a telling point—a point not in any degree compromised by more recent experience—that LSD flashbacks are a rarity and that during such flashbacks violence is all but nonexistent.

Of course, violent and even self-destructive acts are not uncommon in the immediate aftermath of LSD use, while "tripping," as it is colloquially said. The death of Diane Linkletter, daughter of celebrity Art Linkletter, reportedly occurred while she was in the grip of LSD and not during a flashback—although it is difficult to say which is which in the case of an LSD habitué.

Is it possible, then, that in the early-morning hours of Saturday, November 28, 1953, in Room 1018A of the Hotel Statler in New York City, Frank Olson's body and mind were possessed by a dose of LSD he was given that night, which precipitated his exiting the thirteenth-floor window to his death? That possibility is not as fantastical as it might at first blush appear to be. More searching and additional questing lay before me.

ACCORDING TO COLONEL RUWERT'S FURTHER RECOLLECTIONS OF 1953, he journeyed to New York City with Dr. Olson on the Tuesday prior to Olson's death for an emergency visit to Dr. Harold Abramson, a medical doctor who had a grant-in-aid from a CIA cover organization to engage in LSD experimentation, using it as an adjunct to psychotherapy on "abnormal" or mentally disturbed persons. According to Ruwet, Abramson prescribed the use of "sedatives" and a "highball" to relieve Olson's mental distress. Later in the evening of that same Tuesday, Abramson visited Olson's hotel room and "brought a bottle of bourbon and some Nembutal for Dr. Olson." After "a couple of highballs," Abramson left, recommending "to Dr. Olson that he should take a Nembutal which he did at the time and that Dr. Olson take another should he have difficulty sleeping."

Medicating Dr. Olson with two interacting central nervous system depressants had the real potential of killing him, as columnist Dorothy Kilgallen learned to her fatal distress. What sort of medical practitioner would prescribe with such reckless abandon?

We know that Dr. Harold Abramson was a true believer, in the Eric Hoffer sense, in the value of LSD in psychotherapy for disturbed persons. His writing on that theme is bountiful, even up to 1975, when the use of LSD had been long prohibited by federal criminal statutes. It is known also that in 1953 he was under contract with the CIA to experiment with LSD on his mental patients.

Although Dr. Abramson is now deceased, it appeared from the CIA files still available that his associate, Dr. Margaret Ferguson, was still alive in 1993. However, my assistant's phone call to her was abruptly terminated by her. When it was mentioned that the call related to the death of Dr. Olson, she was unwilling to discuss the matter of Dr. Abramson's treatment of Dr. Olson or his experiments with LSD or her knowledge or involvement in it.

The extant record of Dr. Abramson's commitment to LSD in psychotherapy and his ministrations to Dr. Olson tremble with the disquieting possibility that Dr. Abramson might have given Dr. Olson an additional dose of LSD, under a misguided belief that it would relieve his symptoms of depression. That dose might have catapulted him to his death.

For me, however, the likelihood of Dr. Abramson's direct involvement in the death of Dr. Olson is less plausible than that it was the outcome of the CIA's own calculated misdeeds. Not only does the paper record do more than whisper of this possibility, but the statements and silence of persons privy to insider information on the matter who were interviewed or sought to be interviewed by me reinforce that conviction.

I am fully cognizant of the dread scourge of noninvolvement that causes many people to play the clam when information is sought. I am also aware that there are those who refuse to speak because they view the subject under inquiry as too well rung to have any value in pursuing yet another time. Notwithstanding these purported justifications for silence, there are those whose intimate knowledge of critical details imposes upon them a singular duty to speak. Thus, their refusal can legitimately be read as evidence of concealment, and conceivably, even more incriminating implications.

Dr. Margaret Ferguson was not the only one who refused to be interviewed. Vincent Ruwet, both a supervisor of Olson and his supposed friend, also adamantly refused to enter into a dialogue on the death of Dr. Olson. His reiterated one-liner to me during my telephone call was "I have nothing to say on the matter."

I did find people who would talk, though there were major frustrations in the doing.

RETIRED NEW YORK CITY PATROLMAN JOSEPH GUASTAFESTA (MEANing "good feast," as he told me while preening himself, but the correct translation would be "wasted or ruined feast") took a different tack and talked a blue streak, but his uncommunicativeness on Dr. Olson's death was just as indefatigable as that of Colonel Ruwet. He authored the first police report on the death of Dr. Olson and, therefore, presumptively, was the first police officer on the beat to respond to the alarm of a man's having fallen. My letter to the New York City Police Department's pension division seeking to contact him was forwarded to him, but I received no reply from him.

However, I was not to be deterred or flummoxed by this administrative impenetrability. My unflagging investigator, Gary Eldredge, located Guastafesta through driver's license and vehicle registrations as having homes both in New York and in Florida. While attending a scientific convention in Orlando, Florida, I decided to rent a car and pay a surprise visit to the former patrolman at his west coast Florida retreat. As it turned out, my knock at his door went unanswered. The neighbors, however, said Guastafesta was only momentarily absent. So I hid out at a nearby Hooter's, courtesy of a CBS camera crew who were following me about.

Upon my returning to Guastafesta's home in the darkness, I noted a light was on in the dimly lighted interior. I was beginning to sense that I was approaching a fog-laden back alley in a London byway.

With a mike concealed in my pocket connected to the CBS camera crew across the street, I knocked imploringly on Guastafesta's door. "Who's there?" came a gruff and uninviting male voice. "A professor investigating the death of Dr. Frank Olson," I gently responded. Suddenly, a female voice interjected: "I knew they'd find you someday. Don't let him in."

That voice having had its say, the door was opened and Joseph Guastafesta stepped out on the porch, having turned on a bare bulb over my head. As the curtains fluttered on a window next to the front door, Guastafesta shut the door and turned his imposing, full, seventy-year-old frame toward me. "What do you want?" he said with emphasis on each word.

"Your remembrance of Dr. Olson's death in a fall from the Hotel Statler in November 1953 is what brings me to you," I answered in an unquavering voice.

Guastafesta professed to remember little of the events at the time of his investigation into Olson's death. He did confirm night manager Armond Pastore's statement that Olson was still breathing as he lay on the sidewalk. On matters of more vital moment, such as what Lashbrook's statement or behavior might have been when confronted in Room 1018A, his memory went blank and immovably so. As he explained it to me, he was just a "dumb cop."

I cannot comment on that, except to say that his memory was obviously very selective, sometimes showing crystal clarity and sometimes mired in opacity. His failure to recall whether he kept a record of the event in his patrolman's notebook was a challenge to his credibility, as was his memory lapse on whether this was the one and only jumper he had investigated. Inexplicable silence during an interview comes in many tones and hues. A failure to recall is but one dubious kind.

Dr. Robert Lashbrook was to be next in my investigative sights. No devious or long-term tracking was necessitated in this instance. One telephone call said it all. And that all was a most disturbing revelation.

Many persons with insider information whom I contacted were more or less generous with their time. Dr. Lashbrook, then retired from the CIA, permitted a member of my team to question him at some length over the telephone. The most important new and unexpected revelation from Lashbrook was his contradicting all the previous reports of what had caused him to awaken when Olson plummeted to his death.

It was not as the Church Committee and the New York City Police report and all the many other reports, including his own signed report, had said—that "a crash of glass" had cast his sleep aside. It was rather the noise of the window shade spinning in its upper housing. As to whether the glass had been broken at all, he asserted he had no recollection, even though the New York City Police report had simply repeated his own recitals that he had heard "a crash of glass," and he admitted to having gone to the window, put his head outside it, and viewed the body of what he surmised was Dr. Olson on the sidewalk below. In addition, his signed statement to the CIA on December 7, 1953, speaks of Olson having exited through a closed window. Moreover, Lashbrook's remembrance of the window shade's snapping back into its housing was out of joint with Armond Pastore's oft-repeated recollection of the shade's having been pushed through the window's broken glass.

Moreover, my prior interview with Lyman Kirkpatrick, a longtime manufacturer of window shades, left me with little doubt that a shade of the type fitted to the window of Room 1018A would not snap back to its housing if a body had struck it full-tilt.

Was Lashbrook's memory playing tricks on him, or was he playing hob with the truth of the matter? Clearly, his statements must be considered to be the most self-serving of all, since he was the last person known to have seen Dr. Olson before he plunged to his death, and it was he who was charged with supervising Olson's well-being.

I had yet one more interview to pursue, and that was with Dr. Sidney Gottlieb, the man reported to have been instrumental in the death of Patrice Lumumba in Africa. It turned out to be the most mind-bending of all the interviews I had conducted in this investigation.

My meeting with him took place in his home near Culpeper, Virginia, early on a Sunday morning as the sun was peeking over the eastern horizon. Hearing of this impending interview, Eric Olson had cautioned me, a genuine smile creasing his face, not to "drink his coffee." I did and lived to drink more on another day.

I had almost been foiled in this opportunity by my initial attempt to engage him in conversation—unannounced, with CBS cameramen in the background. That encounter came to naught because of his distress over the presence of the media. However, on his initiative in telephoning me at my office, a later meeting was scheduled at which he demanded the right to question me for one hour in return for my interviewing him for an equal period of time—a most uncommon request, but one to which I acceded. His stipulations made me make haste to sit in his presence, the presence, as it appeared, of a most curious man with a puzzling mind.

Even though I had my tape recorder in my pocket at our meeting, I judged it best not to seek my interviewee's permission to turn it on, restricting myself to four pages of handwritten notes and a contemporaneous transcription of them into my tape recorder as I motored home after our meeting.

My overall assessment of this interview was not at all favorable to Dr. Gottlieb, who evidenced more than idle curiosity over my investigation and its findings. His reaction to the glare of the rising sun's blinding me like an interrogation lamp, causing me to seek to move to another chair out of the sun's bright light, had CIA written all over it. Instead of suggesting I move to another chair at the table, he politely insisted that I relocate the same chair upon which I had been sitting up to then. Query: Was the chair "bugged," or was my Robert Redford movie viewer's idea of the CIA verging on paranoia?

Gottlieb's explanations for his actions both before and after Olson's death were at least unsatisfactory and at most incredible. For example, when I asked whether any of the eight unwitting participants in the LSD experiment had been prescreened for any medical disabilities that might put them at risk, he unhesitatingly said, "No." When I inquired whether any medical personnel had been in attendance at the Deep Creek meeting, anticipating that someone of the LSD subjects would have an untoward reaction requiring immediate medical attention, he again replied, "No." Yet his own prior use of LSD, which he openly admitted, had not been so devoid of medical supervision.

Other questions gave rise to answers sufficient to stand plausibility on its head. Had he shredded documents relevant to this matter? "Sure," was his immediate reply.

"But why?" I rejoined.

So that those documents would not be "misunderstood," he answered.

Misunderstood? Or was it, rather, necessary to destroy them so that they would not be understood? I silently mused.

When I asked what action he had first taken upon being apprised by Lashbrook of Olson's death, he said he called his boss, Dr. Gibbons, to arrange a meeting at 5:00 A.M. that very morning. In reply to my raised eyebrows and my inquiry as to the purpose for such a meeting with such urgency, he said the meeting was designed to be "informational."

As he spoke, my mind raced with a stream of unvoiced questions: Informational, and not to plot to cover the tracks of persons responsible for Dr. Olson's death? If informational only, why the rush to have an immediate meeting? If not a deliberate effort to present a united front in the event of a skeptical and official inquiry, why not wait until regular business hours for the "informational" meeting?

Probably the most unsettling, even unnerving, moment in my conversation with Dr. Gottlieb occurred toward its close when he spontaneously sought to enlighten me on a matter of which I might not take due notice—so he thought. He pointedly explained that in 1953 the Russian menace was quite palpable and that it was potentially worsened by the Russians' having cached many kilograms of LSD from the Sandoz laboratory in Switzerland.

Listening awestruck to him as I gazed at a picture of South Africa's Bishop Desmond Tutu on the wall, I was emboldened to ask how he could so recklessly and cavalierly have jeopardized the lives of so many of his own men by the Deep Creek Lodge experiment with LSD.

"Professor," he said without mincing a word, "you just do not understand. I had the security of this country in my hands."

He did not say more, nor need he have done so; nor did I, dumb-founded, offer a rejoinder. The means-end message was pellucidly clear.

Alfred Packer's victims in 1874 John Randolph sketch.

Packer's victims' grave with topsoil removed in 1989.

Packer's victims uncovered in 1989.

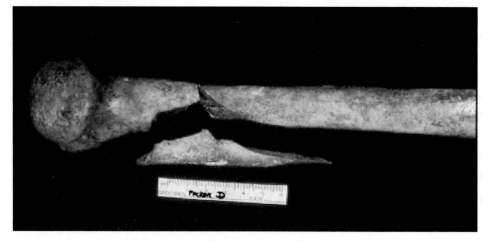

Prospector George "California" Noon's left arm, fractured by
a hatchet blow while he defended himself.

Noon's skull suffered
seven hatchet blows.

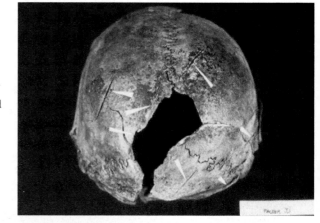

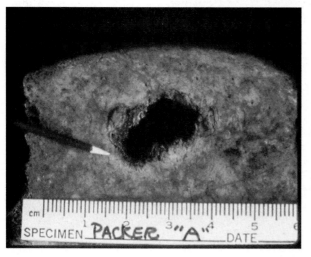

Prospector James Humphrey's
left hip with hole mistaken by
newspapers as a bullet strike.

Carl Weiss in white linen suit in Louisiana state capitol hallway after sustaining multiple bullet wounds.

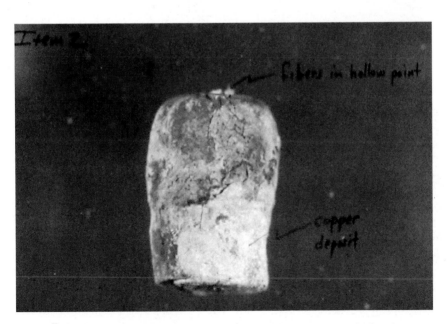

Spent, unexpanded, decelerated hollow-point bullet from Weiss's skull with white linen fibers in its nose. *(Courtesy Luke Haag)*

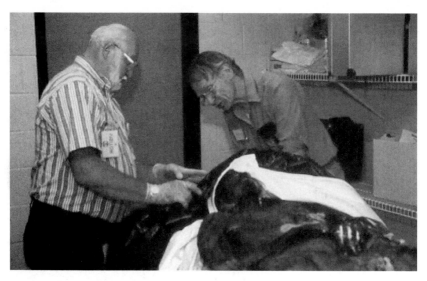

Drs. Jack Frost and Jack Levisky examine Dr. Frank Olson's mummified remains.

Dr. Olson's skull with bloodstaining over left eye from blunt-force injury.

The author explaining Dr. Olson's skull trauma.

The author observes Dan Kysar reconstructing Jesse James's skull.

Jesse's gold-filled teeth, from which his DNA was extracted in 1995.

Facedown outline of Jesse's bones as uncovered in his grave in 1995.

Mary Sullivan, the Boston Strangler's last victim.

Dr. Todd Fenton, in gloves and mask to avoid DNA contamination, examines Sullivan's hyoid bone for a fracture.

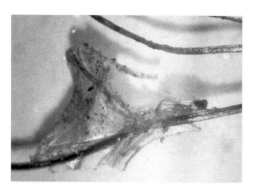

Sullivan's pubic hair with biological material from which the assailant's DNA was identified.

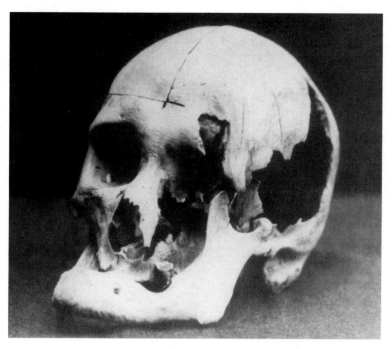

Andrew Borden's skull, as established by its robusticity.
(Courtesy Fall River Historical Society)

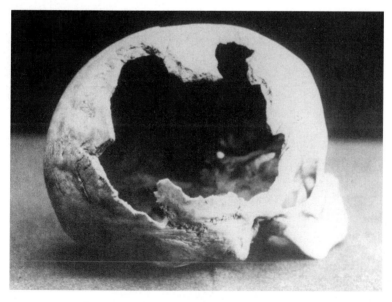

Abby Borden's skull, as established by its gracility.
(Courtesy Fall River Historical Society)

Ground-penetrating radar scanning at Meriwether Lewis's gravesite.

The author loading a .69-caliber gun of a type claimed to be Lewis's suicide weapon. *(Courtesy Luke Haag)*

The author firing the gun to test its muzzle velocity and shot dispersion. *(Courtesy Luke Haag)*

Risking the lives of the unwitting victims of the Deep Creek experiment was simply the necessary means to a greater good—the protection of the national security.

And that was my final interview in this deeply disturbing investigation. I realized then that my professional ivory tower had not equipped me for the CIA's world of feints, flaws, and foibles.

THE ORIGINAL PLAN WAS TO REBURY FRANK OLSON AT MOUNT Olivet Cemetery beside his wife, Alice, in July 1994, but instead his remains were stored for possible reexamination. The bones lay in pieces in boxes under lock and key in my office at GW, and the soft flesh was placed in a vault in storage at York College. These are the places where Olson's remains were deposited for eight years while Eric pursued further avenues of investigation. Not even the presentation of my team's scientific and investigative findings at the National Press Club on November 28, 1994, closed the door to the emergence of new eye-opening details. It was the day following the National Press Club's public disclosure of our work and findings that I received a telephone call from Dr. Robert Gibson, a retired psychiatrist, who indicated that his memory had been jogged by his reading the newspaper accounts of my Olson labors.

Dr. Gibson informed me that he had been the psychiatrist on duty at the Chestnut Lodge Hospital in Rockville, Maryland, when Lashbrook, without identifying himself, had called from New York City just hours before Dr. Olson died. He told me he could never forget that call, although so many years had passed, because the next morning he was the recipient of a second phone call from Dr. Lashbrook stating that Dr. Olson had died in New York City and therefore would not be coming to Chestnut Lodge Hospital.

Dr. Gibson explained that the affair was still vivid in his memory because he grieved over the fact that he had dissuaded Dr. Lashbrook from journeying with Dr. Olson to the hospital on the night of the first phone call. His recommendations against hospitalization had been voiced when Dr. Lashbrook had insisted that Olson was then no danger to himself or to others.

Upon my inquiring about the conversation between him and Dr. Lashbrook on the morning of Olson's death, I was stunned to hear that Lashbrook had given yet another conflicting account of the occurrence. Dr. Gibson affirmed that Lashbrook had said he awakened to find Olson standing in the middle of the hotel room. He had tried to speak to Olson, but Olson had run straight toward the window, through it, and to his death. But Lashbrook had not said a word about whether the window was closed, whether the shade was drawn, or otherwise. However, I now had another conflicting view of the last moments in the life of Frank Olson. And this latest account came from the recollections of an unimpeachable reporter, a psychiatrist of impeccable credentials who had gone on to become the director of the esteemed Sheppard, Enoch and Pratt Hospital in Baltimore, Maryland. This telephone call was memorialized by me in a subsequent taped interview with Dr. Gibson.

That same year, Eric and Nils retained attorney Harry Huge, based in Washington, D.C., to represent them in their effort to reopen the case and instigate a criminal investigation. They had high hopes of suing the government for misrepresentation or concealment of the facts. Huge drafted a fifteen-page memorandum in support of their position. That memorandum, together with the urging of others, persuaded New York's district attorney, Robert Morgenthau, to reopen the investigation. He assigned the case to Stephen Saracco and Daniel Bibb, seasoned prosecutors on the "cold cases" unit. They immediately began to run down leads, and were surprised to learn that the CIA had not turned over to my team Olson's fingerprints or dental records. They also noted my interview with Gottlieb and my unshakable feeling that he was concealing information.

In addition, it was at least odd that CIA Director William Colby had produced documents about Olson's projects that Gottlieb had testified before Congress in 1977 were destroyed. At that time he had asked for immunity, but Saracco was not sure why he would do so, nor why so much material had been redacted from the Olson files. They looked more closely at several aides-de-memoirs that the Olsons had not accessed, believing they held the key to why Frank Olson had been considered a security hazard. One of these memos indicated that Olson had been associated with what it dubbed "un-American" groups. The message was cryptic, but it characterized Frank Olson as having "no inhibitions."

Saracco and Bibb decided to interview William Colby, Vincent Ruwet, Robert Lashbrook, and Sidney Gottlieb. They mailed letters to Ruwet and Colby. Within days, Colby suddenly turned up missing. He left a computer running in his home, a glass of wine on the table, and the lights and radio on. It was a mystery, solved a week later when searchers found his decomposing remains on a Chesapeake Bay island. His death was ruled an accidental drowning when water swamped his canoe.

Saracco and Bibb did manage to meet with Ruwet in Frederick, Maryland, as reported by investigative essayist H. P. Albarelli, and they found him skillfully evasive. They decided to set up a second meeting to explore further what he had said, but on November 16, Vincent Ruwet suffered a fatal heart attack.

Lashbrook, now eighty, refused to respond to the request for an interview, so the DA's office issued a subpoena. Lashbrook refused it. Saracco and Bibb flew to meet him in California, but he would talk only in the presence of an attorney, which resulted in his stonewalling all the questions asked.

Gottlieb was another matter. Even as they were seeking an interview, they learned that he was being sued in civil court by the family of an aspiring artist whom he had allegedly drugged with LSD in Paris in 1952. The incident had made this man unstable and had ruined his life, so it was alleged, possibly precipitating his early death. The situation was consistent with the CIA's surreptitious drugging of hundreds of unwitting American citizens as part of their Cold War experiments. At Senate hearings on the matter in 1977, Gottlieb had stated that these were risks worth taking for the sake of national security, the same position he had expressed to me at our one and only meeting.

With this information in hand, Saracco and Bibb prepared to see him. He had told reporters in the past that he was not going to discuss Frank Olson any further. Nevertheless, the attorneys prepared a draft of a letter to him. On that same day, Gottlieb died from a heart attack.

A bit befuddled and bemused, the prosecutors turned their attention to David Berlin, who had investigated the CIA in 1975 and had brought Frank Olson's LSD dosing to public attention. They wanted to know the details of that investigation. Prophetically enough, before they could make contact with him, Berlin fell in a hotel room and died from head injuries.

Eric Olson, always on the qui vive for new information, discovered that in the assassination training unit of the Israeli Mossad, the Olson case was used as an example of the perfect murder. Eventually, H. P. Albarelli came to Olson in his research into the case. He had contacts among CIA agents in Florida, he said, and some of them had told him they knew who had killed Frank Olson in Room 1018A that night. It was his belief that the perpetrators were contract killers associated with a mob family, and the CIA had hired them to do their dirty work. However, these agents refused to spill the details without the CIA's releasing them from their confidentiality agreements, and that never happened.

Then, in 1997, Eric obtained a copy of the CIA's 1953 assassination manual, now declassified. In it, the scenario promoted for "contrived accident" murder precisely fit like the key that would unlock the door to the mystery of Frank Olson's death—so much so that the only question remaining was whether his murder was the paradigm for the manual or vice versa. That manual clearly suggested disguising a murder as a suicide by dropping the subject from a high window or roof. It also suggested that the perpetrator then play the "horrified witness," and that drugs be used for the subject's preparation. If the subject had to be controlled, blows to the temple or behind the ear were said to be effective means of silencing him.

That manual certainly might explain why the CIA had taken an allegedly psychotic man to a hotel room situated on a high floor rather than to a safe house or hospital. It would also explain the low exit velocity that allowed Olson to hit the wooden impediment on the sidewalk below. That is exactly what would have transpired if he had been dropped rather than propelled through the window.

Through an unnamed source, the DA's office learned that at the Deep Creek meeting, Olson had been given LSD mixed with Meretran, a drug that makes people talk more freely. This confirmed the document from Dr. Abramson, written a few weeks after Olson's death, that the drug had been used to set a trap. He had also told Army researchers that Olson's death had been in the wake of security breaches by him following his 1953 trip overseas, and that only two people knew the complete story about it. One of them had received a substantial sum of money in the month of Olson's death for "services rendered." So had Abramson.

It was learned that an agent who had gone to New York immediately after Olson's death had listened to a conversation between Lashbrook and Abramson that clearly concerned their conspiring to devise a report to the effect that Olson's mental state had been deteriorating. But that was not all. Even more frightful discoveries were in store.

The Olsons discovered documents in Michigan's Gerald Ford Library that indicated the government had been extremely concerned upon learning that Frank Olson's family was questioning the official explanation of his death. The government feared that highly classified information was at stake and that a court might grant full discovery, including information on the nature of Olson's work. That had to be avoided at any cost. As Eric Olson surmised, "In the wake of the Nuremburg trials in the late 1940s, the United States could not afford to be exposed as a sponsor of the sort of research it had prosecuted the Nazis for undertaking."

While these myriad investigations were occurring around me, I was not quiescent. At the urging of Eric and Harry Huge, I took a train from Union Station in Washington, D.C., to Pennsylvania Station in New York City, accompanied all the while by Olson's boxed skull. Traveling with human skeletal remains in tow always leaves me, as Albert Camus said on another subject, "strangely aching." I cannot leave the skull out of sight, but at the same time I cannot treat it in a way to generate suspicions. After all, what would those with prying eyes say to my explanation that I was carrying it to the New York City medical examiner for his inspection?

Upon arriving in New York City, I had my usual professional difficulty of finding my way successfully from Point A, Grand Central Teminal, to Point B, the district attorney's office, where I was to meet a member of the medical examiner's staff. Finding myself lost in the warrens of lower Manhattan, I happened upon a police station, where I asked for directions. The police were forthcoming and very accommodating, to the point of a uniformed officer's volunteering to show me the way, even to the extent of carrying the box with the skull for me. The officer was most courteous and pointed to my being somewhat overburdened, as he said, for a man of my years with the items I was lugging about with me. Needless to say, I declined his offer, not wanting to be mistaken for a follower of Milwaukee's cannibalistic killer Jeffrey Dahmer, who collected the skulls of his victims.

I made my way unescorted to the district attorney's office, where by happenstance I met the author Patricia Cornwell, who said she was there researching material for a book.

"What brought you here?" she asked.

"Oh," I said, "just a skull session with an assistant medical examiner."

She inquired no further, nor did I volunteer anything more.

My meeting with the assistant medical examiner was most unsatisfactory. He entered the meeting with a negative frame of mind, wondering aloud why such a "cold case" would be deserving of his or anyone's attention. It was, therefore, not unexpected that he would downplay the significance of the sizable bloodstain on the skull over Olson's left eye. As he put it, the staining could have come about postmortem. When I asked for his experiential evidence that such a thing had happened in some reported case or cases involving an exhumation many years after the fact, the answer was one that no true scientist would tender: "Everyone knows that to be the case," he said. I could have asked, but, getting the drift of the position he had taken, I pretermitted asking whether he was telling me that bleeding continued even for thirty-six years after death.

Here was a pathologist who wanted a smoking gun, not a skull with a massively bloodstained frontal bone, to induce him to declare homicide the manner of death. I wondered, to myself, with that attitude controlling his decision-making, how many murderers had escaped their just deserts in New York City on his watch.

Foiled and frustrated, I made a quick but courteous departure and in my abstracted frame of mind took a taxi back to Grand Central Station.

Early in 2001, Harry Huge reportedly expected the DA's office to declare Olson's death a murder, which would free the Olsons to file a substantial lawsuit. However, things began to crumble, and by the end of the year Eric discerned that the DA's office had retrenched on their efforts. Notwithstanding, the DA's men had learned that two men had gone into Olson's room the night of his death. Unfortunately, these men were untraceable at this late date. The CIA denied any knowledge of them, but one man appeared to be related to several assassination cases, and therefore no one was going to talk.

I had thought all along of the possibility of the presence of a third person in Room 1018A when Olson exited it. Such a person could have been positioned to enter from the next room through a connecting door. The hotel management had refused my request to disclose the identity of any person resident on November 27, 1953, in the adjoining room. It was said not to be the hotel's policy to reveal the names of its guests, since they might be there under circumstances, shall we say, possibly compromising their marital vows or otherwise.

More was yet in the offing.

In the spring of 2001, Norman Cournoyer, who had been a close friend of Olson's and a colleague at Detrick, contacted the Olson brothers. He had seen the published accounts of the investigations and believed that something was missing. He pointed out that Olson had witnessed harsh interrogations of former Nazis and Soviet spies, and that sometimes these persons had died as a result of those procedures. Olson had heard it said that, despite official denials, the American government had used biological weapons in Korea during the Korean Conflict. Alice Olson had said that Frank, her husband, had expressed distress over this possibility, but she had not known if he actually knew anything with certainty.

On September 12, 2001, Harry Huge suddenly bowed out. He was swamped with too many other cases, he said. Disappointed, Eric found it difficult to find another attorney to take the lead. He decided the time had come to close the case and to rebury his father. He called me in mid-July 2002 to obtain the return of his father's remains. After eight years of Dr. Olson's presence in my office, I parted with his remains with mixed emotions.

On the day before the funeral, Eric and his family held a press conference to call finis to everything he had learned to anyone who was interested. The Olson family had written a twenty-three-page statement for the media, reminding everyone that it had been forty-one years since their father's death, twenty-seven years since the government had offered what they said was the truth, and eight years since the exhumation.

It was their belief, although they had no smoking gun, that their father had been murdered to keep him from disclosing information about a secret CIA program, and that the story of a "bad trip" on LSD had been concocted either as a cover-up for his murder or as a cloak that concealed the dagger they had figuratively used to kill him.

On August 9, 2002, I went in the company of team members Jack Frost and Jack Levisky to Mount Olivet Cemetery on Market Street in Frederick, Maryland, for Dr. Olson's reburial beside his wife Alice. It had been eight years since we had exhumed his remains, and much had transpired in investigations since then. Nils and Eric had discussed with me the possibility of cremating the remains of their father. I prevailed on them not to do so for the sake of Alice, their mother, and for the solace they would take in knowing the investigation was not ended, only interrupted by the reburial.

In Mount Olivet Cemetery, which dates its first interment to 1854, lie hundreds of soldiers from both sides of the Civil War, as well as veterans from the past century's world wars. Eight miles of paved roads run past more than 34,000 graves. Also, an impressive monument to Francis Scott Key, who wrote "The Star-Spangled Banner," stands tall and erect there.

At the funeral, attended by several dozen relatives, friends of the family, and members of my team, the remains were lowered into the ground for what could be their final rest.

From my view and that of the clear majority of my team members, with all the other investigative details, as well as what we found scientifically, Dr. Olson's death was not a suicide. The probabilities, taken together, strongly and relentlessly suggest that it was a homicide.

In the present state of our factual knowledge about the death of Dr. Olson, I would venture to say that the subgaleal hematoma is singular evidence of the possibility that Dr. Olson was struck a stunning blow to the head by some person or instrument prior to his exiting through the window of Room 1018A—a person or persons with a homicidal frame of mind. The convergence of the physical evidence from our scientific investigations with the results of our nonscientific inquiries raises this hypothesis from the merely possible to the realm of real and incontestable probability.

The documentary evidence from 1953 demonstrates a concerted pattern of concealment and deception on the part of those persons and agencies most closely associated with—and most likely to be accountable for—a homicide most foul in the death of Dr. Olson. And the steeled reluctance to be honest, forthright, and candid by persons with knowledge of the occurrence on matters pivotal to the question of homicide or suicide bespeaks an involvement more sinister than mere unconcern, arrogance, or even negligence. The confluence of scientific fact and investigative gleanings points unerringly to the death of Frank Olson as being a homicide, deft, deliberate, and diabolical.

4. JESSE JAMES

The Houdini of Western Outlaws

Jesse James had a wife;
She's a mourner all her life,
His children, they were brave.
Oh, the dirty little coward that shot Mr. Howard!
And they laid Jesse James in his grave.

AUTHOR UNKNOWN
FROM ROBERTUS LOVE, *THE RISE AND FALL OF JESSE JAMES*

No matter how hard we try, no matter how much we wish it were not so, the American outlaw Jesse James just will not go away. From 1865 until 1882, he carved a niche in Missouri history, and over the years since then that niche has expanded to encompass the world. One would think that a murdered figure of such renown, who history tells us is buried in a certain and specific grave, would in fact be in that grave. Yet stories persist that Jesse James died elsewhere than in St. Joseph, Missouri, and that someone else is buried in the grave site in Mount Olivet Cemetery in Kearney, Missouri. It seemed likely that only an exhumation would settle the question, but even that wasn't clear at first. Once we got fully involved, new mysteries unfolded and we learned much more about this American outlaw than just the answer to our original question.

The way I came into the case was quite serendipitous, and the result of some clever maneuvers by my research assistant, Jay Ferguson. He was the politician of the George Washington University law school who knew how to get things done. Once he took an interest, what came next shouldn't have surprised me.

One day in 1994, a man named Emmett Hoctor sent me a nearly undecipherable letter about the Jesse James legends. Jay brought it to my attention, and since I could barely read it, I simply tossed it on the heaping "to do" pile. While I had already exhumed a few historic figures by this time, I wasn't about to respond to everyone who thought he had a story to tell . . . and work for me to do.

But Jay was curious. He asked if he could read the letter and look further into the matter. I told him to be my guest. It wasn't long before he persuaded me to reconsider. He thought it looked like an interesting case. I replied that we would have to find descendants, which would be a time-consuming process, and get their permission to proceed. I didn't think that would be a walk in the park.

But Jay found a direct descendant of Jesse James through Jesse's son, a man in California named Judge James Ross. He found the judge's telephone number and I agreed to call. But the judge would not hear of my exhuming his great-grandfather. He was decidedly uninterested in such a project. I told Jay that was that and the project was dead in the water. Convinced I was right, I went to class. An hour later, upon returning to my office, Jay told me, "It's back on." He explained that he had spoken with Judge Ross himself.

I was surprised. "How did you get him to agree?" I asked.

"I did what you didn't do," he said with a smile. "I told him you would pay for it."

And so that's how the exhumation of Jesse James became a large part of my latter-day professional life, all through the initiative of a curious assistant with the skill and cunning to make things work.

But there was one other relative I still had to sign on to the idea of exhuming Jesse James—a woman in Kansas City, Missouri, named Betty Barr. She is the great-granddaughter of Jesse James through his daughter Mary James. My preliminary plans called for me to visit Kansas City with Dr. George Stephens, an irenic forensic geologist at The George Washington University who has frequently assisted me with soil analysis and other geological features of graves prior to an exhumation.

"Shall we pay Betty Barr a visit?" I asked him.

He was agreeable to giving it a try, she being a close relative of Jesse James with the power to object to any exhumation of his remains. However, when I called her, she was adamant that she wanted nothing to do with such a project and would offer no support. Neither would her mother, who lived with her, agree to an exhumation.

I asked if they would at least allow me to visit with them to plead my case. Ms. Barr begrudgingly agreed but warned me that I would not change their minds. That Sunday morning, just hours before leaving to return to Washington, D.C., I drove with my geologist colleague George Stephens to Ms. Barr's home. She reiterated that she would never allow me to do this. It was, she said, a violation of the sanctity of his grave. I let her speak her mind and then presented my plan.

George and I were there for almost two hours. I did most of the talking, and the questions came to me at a fast and furious pace. By the time we left, we not only had permission to proceed from Betty Barr and her mother, but they were eventually to become staunch advocates of what we were doing for the memory of their forebear.

As we drove away, George turned to me and quietly remarked, "Jim, you have a silver tongue."

Although my support team of volunteer scientists was in place, we still had to do much aboveground digging into the life and times of Jesse James. A better understanding was imperative for us of the century-old controversy over whether he had been killed in St. Joseph, Missouri, on April 3, 1882, or had faked his death to escape his pursuers.

AMERICA AT ITS BEST AND AMERICA AT ITS WORST—MOST OF ALL at its worst. That sums up the life and the lore of Jesse James, Missouri's homespun folk hero, whose Paul Bunyanesque exploits keep enlarging year after year. Inscrutable paradoxes often mark those to whom we ascribe the status of notables, and no more is this in evidence than in our attitude toward Jesse James.

Even though he was never arrested or convicted of any crime, it is still generally conceded that Jesse James, with his brother Frank and their gang, roamed, ravaged, and robbed across a vast midwestern territory during the better part of twenty years. Neither fingerprinting nor Bertillonage (identification by bodily measurement) was in use for personal identification at the time, but that does not alone explain the inability of law enforcement to stop Jesse James. Neither the railroad company that he had victimized nor the reckless resourcefulness of the Pinkertons was up to the task. It seems self-evident that Jesse and his gang could not have outclassed or outlasted their pursuers for so long a time without the unremitting support of the law-abiding private citizenry. If necessarily so, then the James boys could not have been all bad—or at least they must have had qualities considered by their backers worth keeping unharnessed.

Pitted against the banks and railroads as they were, the James gang gave violent voice to those who lacked the nerve to speak out in opposition to the behemoths of industrial America. More than that, their lifestyles of hardy independence, a trait that had characterized the pioneer spirit in settling the American West, was thought to be in serious jeopardy. Add to these concerns the way the James gang symbolized the outrage of Southerners and Southern sympathizers at their defeat in the Civil War and you have more than enough to explain the numbers of people who sheltered the James boys for so long. Despite being outlaws, they were also heroes to many, and mythical stories inevitably grew up around them.

These Homeric tales told of Jesse's bold daredeviltry and his self-sacrificing compassion shower him with praise for successfully playing to the hilt the role of defender of the poor and the defenseless. Yet this publicly anointed Robin Hood of the West was in reality just another "social bandit." It was irrelevant to his status as a folk hero whether or not he performed the good deeds ascribed to him. He was perceived as a Tommy-good-guns who kept the needs of the oppressed in the forefront of his actions.

The tale, more than any other, that has given enduring support to Jesse's chivalrous generosity recounts Jesse's giving aid to a widow in her hour of greatest need. What follows is the most popularized version of this story.

It was at dawn's first light when a venal banker arrived on horseback at an impoverished widow's home where Jesse and his brother Frank had been given food and housing for the night. "Pay the $510 that you owe me on your mortgage or leave at once," the banker pompously threatened, a smirk of odious pleasure darkening his face. Before the cowed widow could reply, Jesse, with Frank at his side, stepped out of the shadow of the doorway of the cottage. The banker, taken aback, watched as the outlaw counted $510 out of a much larger roll of bills cradled in his hand.

"Here's your money, you craven widow-sucker," Jesse hissed as he handed over the money. Thoroughly dismantled, the banker could only take the bills and leave, but not without a parting shot to the widow. "I'll be back another day," he warned. "Mark my words." And off he trotted.

No sooner had the greedy moneylender left than Jesse and Frank saddled their horses and bid farewell to the widow, whose outpourings of gratitude followed them as they rode off. Making tracks at a fast clip, they waylaid the banker before he had reached the security of the nearby town. Quickly relieving the furious man of the $510 they had given him, plus other valuables, Jesse and Frank rode off into the daylight of the mythmakers.

Inevitably, there were those who would credit this "poor widow" story as fact. In his early biography of Jesse James, journalist Robertus Love considered that tale to have more than a glint of truth. Yet the exact day or even year of the occurrence, not to mention the identity of the widow or the location of her mortgaged home, is shrouded in mysterious uncertainty. Fables are spun out of far less than scientific data.

Yet it was not only Jesse James's benevolence that was widely touted but also his unmatched skill with guns, making it inevitable that his gunslinging prowess would be compared to the equally legendary Wild Bill Hickok. Both were traveling men, casting their nets over a wide geographical area, and inexorably the stories about them coalesced. There was more of Missouri than Texas in Jesse James, just as there was more of Texas than Missouri in Wild Bill Hickok. But otherwise the two were equally prepossessing as gunslingers—that is, until in their wanderings they "chanced" to meet, according to the well-worn tale.

It happened that Wild Bill arrived in a Texas cow town, where he was unsure as to whether his deadly reputation as a gunfighter had preceded him. After observing the presence of a considerable number of quick-draw artists whose marksmanship seemed unapproachable, Wild Bill decided to stake his claim to the title of the fastest and the most accurate gun in town. Spying a lone, rail-thin tree standing some fifty yards away and noticing that a quiet man sporting a wry half-smile, half-sneer was watching his actions, Wild Bill whipped out his six-shooter. With the speed of an express train rollicking though a rural way station, he fired. The bullet sped to its target and planted itself smack in the center of the spindly target.

"By Jehovah," Wild Bill proudly exclaimed, "with shooting like that I could cut a man's hair without touching his head." With this boast still on his lips, he turned to the quiet onlooker. The stranger threw his hands to his sides, drew his pistols, pointed them at the same tree Wild Bill's bullet had struck, and fired both pistols simultaneously. The bullets sped to their target, one landing above Wild Bill's bullet, hugging it from the distance of a hairbreadth, the other hitting home a tad below.

Speaking through tight lips and narrowed eyes, the stranger murmured, "Better open your hair-splitting pistol parlor pretty quick, for Jesse James is likely to be there parting hairs a mite before you." According to the tale, Wild Bill just shrugged his shoulders in defeat and made haste to leave town, conceding that he was no match for the marksmanship of Jesse James.

Much of the fascination, indeed the captivation, with Jesse James over the years must be credited to the circumstances of his death. Unlike Jesse's older brother, Frank, whose name obviously lacked the same alliterative sparkle and the Old Testament vintage of Jesse's, tradition tells us that Jesse's worst enemy was an enemy in the ranks of his own gang—a man named Bob Ford. Whereas Frank James eventually surrendered and was tried and acquitted for his part in the Winston, Missouri, train robbery (living on to a robust old age as gatekeeper for paying tourists at the James farm), Jesse James was cut down in his prime. The standard story of his death has it that on the morning of April 3, 1882, Jesse, thirty-four years old, was straightening (or dusting) a picture in the parlor of his home in St. Joseph, Missouri, when Bob Ford fired a single bullet at him from behind and killed him.

Jesse's death conceivably rounded out the paradoxes of his life. A desperado on the run, who for many years cannily eluded law enforcement, was brought low by his faith in an untried newcomer to his band. A man who generally trusted only side arms chanced that day to separate himself from them and be vulnerable. Jesse James let his guard down this once, and he never again had another opportunity to do so.

THE FIRST REPORT WE HAVE OF THE DEATH OF JESSE JAMES COMES from a St. Joseph news "extra," dated April 3, 1882, and titled "Jesse! By Jehovah!" I paraphrase it as follows:

On April 3, 1882, Jesse James, the Missouri outlaw, before whom the deeds of Fra Diavolo, Dick Turpin, and Schinderhannes dwindled into insignificance, was instantly killed by a boy twenty years old named Robert Ford, at James's temporary residence on the corner of Thirteenth and Lafayette Streets in St. Joseph, Missouri. The state had offered a $10,000 reward for the body of the brigand, dead or alive. (Official proof is lacking that the full reward was ever paid to either Bob or his brother, Charley. James historians agree on the fact that considerably less than the $10,000 reward was ultimately paid.)

Jesse, with his wife and two children, had come to St. Joseph in November 1881 accompanied by Charley Ford. Under the name of Thomas Howard, Jesse secured a white one-story cottage with green shutters at 1381 Lafayette Street for $14 a month. During the last week of March in 1882, Robert Ford had joined up with Jesse.

Jesse's plan was to rob the bank at nearby Platte City, because a murder trial was to commence there and he knew the citizens would be distracted by their attention to it. Charley suggested his brother Robert as a gang member. Jesse had met the boy three years earlier and consented to give him a shot. Jesse, Charley, and Bob Ford remained in Jesse's house together for a week, plotting the right time to make the raid. All the while, the brothers were watching for an opportunity to shoot Jesse for the reward, but Jesse was always wary and armed.

Then, unexpectedly, their opportunity presented itself.

On the morning of April 3, breakfast at the James residence was over. Charley and Jesse had been in the stable, currying their horses. On returning to the parlor where Robert Ford was waiting, Jesse supposedly said, "It's an awfully hot day." He pulled off his coat and vest and tossed them on the bed. Then he added, "I guess I'll take off my pistols, for fear somebody will see them if I walk in the yard."

He unbuckled the belt in which he carried two .45-caliber revolvers, one a Smith & Wesson and the other a Colt, and laid them on the bed. He then picked up a dusting brush to dust off some pictures that hung on the wall. (Many latter-day historians refuse to believe that a real man of Jesse's ilk would deign to dust—a distinctly womanly task, so it is said. Instead they have Jesse straightening a picture, a task that even a "macho" person might not find below his leonine dignity.)

To perform this simple task, Jesse climbed up onto a chair, so it is said. His back was now turned to the Ford brothers, who silently stepped between Jesse and his revolvers. At a motion from Charley, both drew their guns. Robert was the quicker of the two, and in one motion he had the gun's muzzle close to the back of the outlaw's head.

Yet Jesse had acute hearing. "He made a motion as if to turn his head to ascertain the cause of that suspicious sound," the newspaper reads, "but too late. A nervous pressure on the trigger, a quick flash, a sharp report, and the well-directed ball crashed through the outlaw's skull."

He didn't cry out. He just took the fatal shot and fell heavily backward upon the carpeted floor. The ball, as the account has it, entered the base of the skull and made its way out through his forehead, over the left eye. It had been fired out of a Colt .45, improved-pattern, silver-mounted and pearl-handled pistol—a weapon that Jesse had presented to his slayer only a few days earlier.

Jesse's wife, Zee, was in the kitchen. She heard the shot and came running into the front room. She saw her husband lying on the wooden floor and his killers, each holding a revolver in his hand, making for the fence in the rear of the property. Robert had reached the enclosure and was in the act of scaling it when she stepped to the door and called to him: "Robert, you have done this! Come back!"

He reportedly answered, "I swear to God I didn't!"

The Fords then returned to the house while Mrs. James ran to her husband's side and lifted up his head. He was still alive, and it seemed to her that he wanted to say something but could not. She tried to wash away the blood that was coursing over his face from the hole in his forehead, but it seemed to her that it came faster than she could wipe it away. Then, cradled in her arms, the notorious outlaw Jesse James passed to his eternal reward, if such it was to be.

Charley Ford explained to her that "a pistol had accidentally gone off."

When Zee indicated her disbelief in their explanation, the brothers left to publicize their perfidy. They went to the telegraph office and sent a message to Sheriff Timberlake of Clay County and to Governor Crittenden of their killing Jesse James. Then they surrendered to Marshal Enos Craig.

Later that day, in the presence of an immense crowd, an inquest was reportedly held. Mrs. James accompanied the officers to the city hall for this proceeding, having previously left her two children—Jesse, aged seven, and Mary, aged three—with a neighbor.

The report of the killing of the notorious outlaw spread like wildfire, and as usual, the description assumed every variety of form and lack of substance. Very few credited the reports, however, simply laughing at the idea that Jesse James, the West's elusive Pimpernel, could have been gunned down in such an unprotected fashion.

Nevertheless, excitement ran high, and when one confirming report succeeded another, crowds gathered at the Seidenfaden undertaking establishment where Jesse had been taken that morning. People also showed up at the city hall and on every street corner. The event, whether it had occurred or not, was a humdinger of a story about which the people were all abuzz.

The body, which lay in a remote room of the city morgue, was taken out of the casket and placed upon a table. The features appeared natural but were reportedly disfigured by the bloody hole over the left eye. Then the corpse was cleanly dressed and a photograph taken. Jesse's large, cavernous eyes were closed, as if he were in a calm slumber. As a contemporary newspaper reported: "Only the lower part of the face, the square cheek bones, the stout, prominent chin, covered with a soft, sandy beard, and the thin, firmly-closed lips, in a measure betrayed the determined will and iron courage of the dead."

An inspection of the body, according to this news report, revealed two large bullet holes from previous shootings on the right side of the breast, within three inches of the nipple, a bullet wound in the leg, and the absence of the tip of the middle finger of the left hand. These identifying features have become the marks most associated with the real and true Jesse James.

When the body was released for burial, it was packed in ice and transported from St. Joseph by train to the James farm in Kearney, Missouri, fifteen miles northeast of Kansas City, Missouri. Jesse's mother, Zerelda James Samuel, being fearful of her son's grave being despoiled, had him buried at his family homestead in Kearney, positioned in the front yard so that she could keep watch over it night and day.

Twenty years later, Zerelda James considered putting the James farm up for sale. Consequently, on June 29, 1902, in the presence of Jesse Edwards James (Jesse's surviving son) and an assemblage of notables from the gun-toting years of Jesse's prime (but in the absence of the ailing Frank James), the remains from the James farm were disinterred and transferred to their present burial place at Mount Olivet Cemetery in Kearney, Missouri. That grave is marked "Jesse W. James."

DID BOB FORD REALLY KILL THE NOTORIOUS OUTLAW JESSE Woodson James on April 3, 1882, in his house high atop Lafayette Street in St. Joseph, Missouri? If not Jesse, then who met his maker, courtesy of Bob Ford's revolver, on that fateful day?

Since the reported behavior that made his killing possible was so out of character for Jesse James (he had eluded law enforcement and the Pinkertons for some sixteen years), there were those, many indeed, who would not admit that his death could so simply and so readily have been brought to pass. For this and other reasons have the legends of Jesse James's survival emerged and reemerged over the years as the most durable hallmark of the James folklore.

Some say the assassinated man was an unwitting stand-in by the name of Charlie Bigelow. Others announce that there were in truth two Jesse Jameses: one, the true Jesse Woodson James, and the other, Jesse R. James, who was a Jesse Woodson James look-alike who could fool even Jesse's older brother Frank. In life and in death, he passed himself off, it is said, as the authentic Jesse.

It must be remembered that when it comes to tall tales, the life of Jesse James was just brimming over with the tallest of the tall. Should any the less be expected with respect to his death?

In the history of the world there have been innumerable persons who are purported to have escaped death like Jesse James. Billy the Kid; John Wilkes Booth; the Dauphin, son of Marie Antoinette; and Anastasia, daughter of Czar Nicholas II are the most prominent among these "escape artists," along with Jesse James. Some had stand-ins who took death's licks for them. Others just evaded their fate through personal conniving or with the aid of conspiratorial allies. Still others make no pretense of explaining their reappearance after their reported death. Apparently, they hearken to the words of Mark Twain that the reports of their deaths were grossly exaggerated.

Society at large takes it as axiomatic that notable persons and those with a simulated veneer of fame never die in ignoble ways. The public's preference is to treat these notables as having escaped a death that would otherwise debase their dignity as imperishable icons. In point of fact, there may not have been a single person of importance in history who after death has not, at one time or another, been said to have escaped such an end. I call them "escapists."

Some of those alleged to have survived are never really tracked to any actual living person. Marilyn Monroe's suicide is cast aside in favor of the notion that she has been kept incommunicado in an insane asylum by dint of a grand government conspiracy to prevent her from pointing the finger of extramarital guilt at certain persons of prominence in our country's governmental hierarchy. The actor James Dean, it is contended, is similarly living out his days in an asylum to hide the tragic disfigurement resulting from his 1955 car accident.

However, no one can match, in numbers alone, the more than twenty doppelgängers who have crowded the stage proclaiming their right to be designated as the one and the only Jesse James. One of the two major sources of tourist interest in Granbury, Texas, for example, is the grave of J. Frank Dalton, with a headstone describing him as Jesse James. (The other is the alleged fact that the escapist John Wilkes Booth, posing as John St. Helen, lighted on Granbury before moving on to commit suicide in Enid, Oklahoma.)

All escapists have one feature in common: Not one—Elvis Presley included—has ever been proved with any degree of acceptable certainty to have actually performed that death-defying feat.

SINCE THE MANY CONFLICTING ACCOUNTS OF JESSE'S DEATH BELIE our ability to know the historical truth about this killing, I had to consider several facts already known about the incident and its aftermath, as well as develop a context for the activities of my forensic team. The first complication was that James had been moved from his original grave.

The person shot in St. Joseph was first buried in Kearney, Missouri, on the James farm. Then, twenty years later in 1902 when his mother, Zerelda, wanted him moved, he was exhumed and reburied in Mount Olivet Cemetery in Kearney. In 1978, another excavation took place at the original burial place on the farm. This dig unearthed a spent bullet and an assortment of small bone fragments, some from animals and others left behind in the 1902 exhumation, as well as multiple hair strands. The hair and bullet went into the museum in operation at the James farm, but the bone fragments, after analysis, were reburied in a plastic container.

Thus, there were two grave sites, as well as several accounts of what had actually happened to Jesse James that fretful April day in 1882. In addition, we were faced with a dispute over whether the bullet that killed Jesse James had actually exited from his head and lodged in the wall of his home. And there was one more mystery awaiting us but which was to appear only after we had exhumed his remains.

A human skull, having been struck by a bullet, can be most informative concerning the event. The distance from which the bullet was fired at the skull can be approximated scientifically, based on the existence of traces of black powder from the propellant, as well as from the wound signature on the skull. Contact bullet wounds have a characteristic pattern. So if Bob Ford had held his pistol against the rear of Jesse's head, that fact would be determinable.

The entrance site can also give some indication of the bullet's trajectory as it entered the skull. This is a feature of what is known as external ballistics. If Jesse James was standing on a chair, dusting or adjusting a picture, and his assailant shot upward while standing on the floor behind him, then the angle of entry should, upon reconstruction of the skull, be compatible with an upward shooting. If not, then the traditional picture of his being killed by an upward shooting from the rear by Bob Ford would be suspect.

As to the dispute over whether the projectile that Bob Ford fired actually went through Jesse James's head (known to forensic pathologists as creating a perforating wound) or entered it and did not exit (known as a penetrating wound), the details were shrouded in ambiguity. In support of the view that the bullet exited is the existence of a hole in the living room wall in the St. Joseph house where the killing occurred. Further, pictures of Jesse James in death show what appears to be a linear laceration, with a vertical orientation, next to his left eye, and blood on the shirt covering his left shoulder. Yet no bullet was ever said to have been recovered in the home and the mark on Jesse's forehead could just as well have been a bloody laceration from falling to the floor and hitting his head, causing the apparent bloodstain on his shirt.

According to newspaper reports, Bob Ford gave testimony at the inquest into Jesse's death that the bullet fired from his gun had in fact exited from Jesse's skull. In support of Ford's recollection, there is an account in the *St. Louis Post-Dispatch* of June 29, 1902, penned by noted James author Robertus Love, to the effect that during the course of James's exhumation in 1902, his skull was found to have a bullet hole in the center of the forehead. It also was said that Jesse James's son, in handling his father's skull in 1902, saw a quarter-sized hole in its left rear. But if the bullet emerged from Jesse's forehead, then medically speaking his left eye should have hemorrhaged to give the appearance of a raccoon's eye, which the coroner should have observed. Nothing of so evident a characteristic was ever noted.

After the 1978 analysis of the bone fragments discovered at the James farm, Dr. Michael Finnegan's published report indicated that he examined, both microscopically and radiographically, a small portion of the bone from "the base of the occiput, apparently from the right side." It could not be said conclusively that this bone fragment had been broken at the time of death, he said, as opposed to being a postmortem artifact. Yet the fact that the X-ray examination in 1978 of this fragment did not disclose any "lead or heavy metal fragmentation that would suggest a gunshot wound" gave rise to the possibility that the fragment was not from the bullet entry site at the base of Jesse's skull.

My preliminary investigations were further complicated by a report dated January 2, 1979, from a firearms examiner with the Regional Criminalistics Laboratory in Missouri, which commented on an examination of a .38-caliber bullet, possibly fired by a Smith & Wesson, found at the excavated site at the James farm. Was this the same bullet that was seen at the autopsy in 1882 to be partially embedded in the bone of Jesse's intracranial vault? If so, why wasn't it removed at the autopsy? If not, then this .38-caliber bullet was probably an antemortem artifact that Jesse had carried in life and to his grave.

Whether the bullet entering Jesse's skull was fired from a .45-caliber Colt or other handgun is also a matter of considerable debate. Many persons believe the death-dealing weapon was a .44-caliber Smith & Wesson. Finis C. Farr, private secretary to Governor Thomas T. Crittenden, Jesse James's most pronounced antagonist, wrote in 1902 that Bob Ford stated that the weapon he used to kill Jesse James was a "Colt .44."

However, the possibility that it could have been some other pistol, such as a .38-caliber Smith & Wesson, cannot be entirely discounted. This is made particularly plain from the 1978 excavation's finding of a .38-caliber spent bullet with the typical appearance of a deformed bullet that had struck bone.

As to this .38 bullet, there are at least four theories in explanation of its origin:

1. It was a bullet Jesse James had carried with him from a prior wounding. Some say it was received in the ill-starred Northfield raid. That is highly unlikely, however, if only for the reason that the manufacture and then distribution out of Chicago, Illinois, of this improved .38-caliber cartridge occurred just a few months before the Northfield raid. It would be an unusually rapid distribution that would have enabled that cartridge to have been brought into action at Northfield, Minnesota.

2. It was a second bullet striking Jesse at the time of his being shot by Bob Ford, possibly fired by Charley Ford.

3. It was the bullet that killed Jesse James. If so, then either Bob Ford discharged it or his older brother Charley Ford did so, they being the only two persons in the living room with Jesse at the time of his death.
4. It was a bullet received by Jesse when he attempted to surrender to the Union forces during the Civil War.

Even the James Farm Museum staff seemed confused on the question of what weapon was the murder weapon. The exhibits displayed upon my latest visit included a .45-caliber Colt revolver, said to be identical to the weapon used to kill Jesse James. On an opposite wall a .44-caliber Smith & Wesson is exhibited and said to be a "sister" weapon to the .45-caliber Colt. But a .44-caliber Smith & Wesson is not, by a long shot, a match or a "sister" to a .45-caliber Colt.

In addition, despite some accounts to the contrary, Charley Ford is routinely thought not to have fired off any rounds.

Even though the alleged text of the 1882 inquest adopts the report of Charley Ford's firing, there is no official record in the Buchanan Circuit Court (St. Joseph, Missouri) of his inquest or its text ever having been filed or retained there.

Added to the mysteries of the actual shooting incident are the rumors and tales that the outlaw Jesse James was not the man shot. So not only did we, in our preliminary investigations, face problems with inconsistencies and unanswered questions in the official and unofficial reports, we also faced a major issue relating to the identity of the remains buried at Mount Olivet Cemetery. Thus, our task had many diverse and puzzling dimensions, finding the answers to which would involve the labor and the talents of many different scientific experts.

THE MAJORITY OF HISTORIANS BELIEVE THAT JESSE JAMES DIED ON April 3, 1882, having been shot by Bob Ford in St. Joseph, Missouri. However, some present a contrary account of the events of that day, including one in which Jesse faked his death in order to escape those hunting him and another in which a man named Charlie Bigelow took the bullet intended for Jesse. The Charlie Bigelow impersonation story is said by author Ted Yeatman to be "a fixture in James-imposter lore."

Bigelow was reputed to have looked astonishingly like James and was said to have claimed on several occasions prior to 1882 that he was, in fact, Jesse James. So it was Bigelow who died in St. Joseph, and the surviving Jesse James is purported to have sung in the choir at his own (really Bigelow's) funeral and, by some accounts, had taken part as the mysterious sixth pallbearer at the bogus funeral. Jesse is then said to have departed for South America, later returning to the States, where he lived out his life under various aliases.

Most accounts of post-1882 Jesse James sightings focus on J. Frank Dalton of Lawton, Oklahoma, who at age one hundred announced to the startled world that he was really Jesse James. He stated that he hadn't revealed his true identity until then because he, along with the others involved in the cover-up, had vowed to keep it secret until their hundredth birthdays. Dalton bore a strong physical resemblance to James and had several of James's distinguishing characteristics, such as a damaged left index finger. (However, it should be noted that, in June 1864, Jesse's left middle finger, not his index finger, was shot off.) Dalton also had bullet wounds on his forehead and chest, and had petitioned to have his name legally changed back to Jesse James, but the court in Franklin County, Missouri, had denied his request. Yet Dalton had the final say, of sorts, for he had "Jesse James" inscribed on his tombstone.

The Jesse-did-not-die faction supports its claims by questioning many of the events surrounding Jesse's death. First, they point out that Jesse James would never have taken off his guns in broad daylight, especially since he feared he was being watched. Also, they question why he would have had to stand on a chair to dust or adjust a picture in his living room since he was nearly six feet tall and in those days most pictures were hung at a low height, since ceilings were usually lower than they are today. In addition, when first shown the body of the dead man, Zerelda Samuel, James's mother, is reputed to have initially exclaimed, "No, gentlemen, that is not my son." Later she reversed her story, but the Jesse-lived-on adherents argue that her change of mind is plausible only as an effort to gain peace and quiet for the fugitive by misleading the rest of the world into believing he was indeed dead.

Moreover, since no one of the James family, including his companion-in-crime and brother, Frank, sought to avenge Jesse's death (the dog-that-did-not-bark canard is at work here), it is said to be unlikely that he did in fact die in 1882. While the *Kansas City Times* identified Jesse's body from a photograph, that identification has been questioned, since only two photographs of James were then known to exist. Furthermore, the Dalton faction seeks to bolster their thesis by noting the fact that faking funerals was a common practice in the post–Civil War South. They suggest that western desperado Bloody Bill Anderson and Lincoln assassin John Wilkes Booth both faked their funerals. So for Jesse James to have fabricated his own funeral would have been part of a well-entrenched pattern.

The most incredible and outlandish James sighting, however, was a 1993 report in the grocery store tabloid *The Sun* of his having robbed a bank at age 121. (Elvis Presley was probably in the bank at the time, for one fairy tale deserves another.) An unidentified forensic anthropologist, according to *The Sun* report, compared the bank security camera photo of the robber with earlier pictures of James and concluded that it was definitely Jesse James.

In sum, the many and varied stories of Jesse James's death or his escape from death have given rise to oft-repeated claims from many persons that they are descended from the Jesse James whom Bob Ford did not kill. While most historians, along with the known descendants of the James family, roundly denounce these claims, much ink has been spilled in support of the contrary view, causing unrelieved anguish to the true relatives of Jesse James.

These diverse and often conflicting assertions have so rankled the clearly authentic descendants—mainly of Jesse's two children and his sister, Susan Lavenia—that over these many years since 1882 they have diligently tried to put an end to the matter of his death. They have even proceeded with litigation to confirm their genealogical status against others' claims. But nothing has stilled the strident voices of those who maintain that Robert Ford did not murder Jesse James, and that consequently they are his kin. We had our share of that as well.

LITTLE WAS IT REALIZED AT THE INCEPTION OF THIS SCIENTIFIC investigation that we would be inundated with grandmother stories. Yes, there were also tales traceable to grandfathers, but by far the largest number of telephone calls and letters were from persons who insisted that their grandmothers had told them thus and so. And, naturally, we were expected to take these remembrances as entirely reliable. After all, grandmothers do not lie.

Far from distracting us from our researches and investigations, the contacts we had with enthusiastic persons who had tales to tell were convincing proof that the mystery of whether Jesse James died in 1882 was alive and well, and it needed to be put to the scientific test.

A sizable number of the callers expressed the conviction that Bob Ford did not kill Jesse in 1882. One woman informed us that her grandmother had played with Frank and Jesse James when she was a child. (We made no attempt to delve into the obvious disparity in the ages of the persons involved.) Another person recalled her grandmother describing Jesse traipsing about in ill-concealed anonymity in Kansas City many years after 1882. The writer staunchly believed the truth of her grandmother's recollection since her grandmother was a "no nonsense" person.

One caller remembered his grandmother's having said that Jesse was hanged in New Mexico in 1913. Another said that Jesse assumed the name "Bobo" and told wild stories of the Old West to the amusement of the town's children.

The true descendants of Jesse James stand in wonderment at all of this. Why would anyone, they ask, want to be related to a desperado of such epic proportions?

Exactly, is the only rational reply. But not reason, only science, could put these vagrant apparitions to rest.

The primary objective of this exhumation project was to identify the remains said to be in the grave of Jesse W. James as precisely as science would permit. Our plan was to determine this linkage through DNA (nuclear and/or mitochondrial) analysis, computerized photographic skull superimposition, anthropological analyses, and other scientific means.

That meant recruiting team members with expertise in molecular biology (DNA analysis), anthropology, and computer technology. We also had need of geologists, firearms specialists, a radiographer, and a forensic pathologist, among sundry other scientific disciplines.

The problem we faced from the start was how to deal with the large number of people who believed or wished themselves to be related to Jesse James, no matter how distant or tenuous the connection. The settled genealogy of the Jesse James family, founded on the firm conviction that Jesse James was married to Zerelda "Zee" Mimms, his first cousin, is that he was survived by his two children, Jesse and Mary, as well as a brother, Frank, and a sister, Susan.

With a traceable genealogy to living descendants, we hoped to use DNA analyses to make a definitive determination. DNA, the genetic blueprint for all living things, comes in two types in human beings. Nuclear DNA is found within the nucleus of all nucleated cells, such as white blood cells or sperm cells. Multiple copies of mitochondrial DNA (mtDNA) are located in the cytoplasm of cells.

Nuclear DNA has great value in matching blood or other bodily tissues from a crime scene to a particular person. It is also significant in solving questions of paternity, but it is of no value, statistically speaking, in tracking familial relationships for persons related beyond that of parent and child. That's why the use of mtDNA comes into necessary prominence.

Every person's mtDNA is inherited entirely from his or her mother. Thus, the matching of one person's mtDNA to that of another requires that they be linked matrilineally from mother to daughter to daughter and ultimately to living persons, whether male or female. Applying this knowledge to the question of the identity of the remains marked as Jesse James in Mount Olivet Cemetery, it was essential for our purposes to locate living relatives whose mtDNA was incontrovertibly the same as that of Jesse James by reason of their provable relation to James's mother, Zerelda.

In other words, the mtDNA of direct descendants of Jesse James would not suffice. It was his sister, Susan, and her descendants whose mtDNA would match that of Jesse James, they, like Jesse, bearing the mtDNA of Zerelda James, Jesse and Susan's mother.

The first order of the DNA business, then, was finding Susan's living relatives who were descended in an unbroken matrilineal line from her. According to a suggestion I received from Judge James Ross, Jesse James's great-grandson, finding a known mtDNA sample would be as simple as simple could be. He recommended and offered his consent to an exhumation of Zerelda James, Jesse's mother, whose grave adjoins that of her son. Not anxious to conduct more exhumations than were imperative for my scientific purposes, I demurred at his proposal.

"Unfortunately," I informed His Honor, "the law of Missouri will not permit an exhumation of Zerelda."

"Why is that?" Judge Ross inquired with an air of judicial suspicion.

"Because Zerelda died a natural death," I dissembled, "while the death of her son has been rife with doubt and conflict as to its occurrence and cause. There is, therefore, no allowable basis for exhuming someone in Missouri who dies a natural death." I knew full well that the local medical examiner had the statutory power to request an exhumation even if the death had been previously classified as a natural one. I just was not of a mind to add one more exhumation to my agenda, unless out of dire necessity.

Judge Ross accepted my explanation without my citing the pertinent statutory references and without objection, something of a rarity in my encounters with judges in connection with exhumations. Indeed, I say that even though I had not yet confronted the Federal District Court judge in Nashville, Tennessee, who single-handedly dashed my hopes for the proposed exhumation of famed explorer Meriwether Lewis. But more on that subject in a later chapter.

With Zerelda's exhumation temporarily out of the mtDNA picture, I continued the task of tracing the line of female descent from Susan James. With the invaluable and able aid of Philip Steele and Gary Chilcote, both friendly, notable, and knowledgeable Jesse James historians, I located Robert Jackson, Esq., Susan James's great-grandson, and his nephew, Mark Nichols—two living descendants from Oklahoma who satisfied the mtDNA requirement of being matrilineally connected to Susan James.

Both Attorney Jackson and Nichols graciously and immediately consented to provide a specimen of their blood. In those days of DNA profiling, venous blood was necessary for DNA typing. Today a buccal swab for cells from the mouth will suffice. With these samples, I could compare their mtDNA with that from the grave of the supposed Jesse James, assuming we had no other alternative and that we obtained legal authority for an exhumation.

But the mtDNA of Jackson and Nichols was only one half of the matching equation. For our comparison purposes, we also required a questioned sample of Jesse's mtDNA to compare to that of his sister Susan's descendants. The only available specimens that might contain James's mtDNA outside the Mount Olivet grave were strands of hair recovered in 1978 at the site of his original burial. The hair in question was in the custody of the James farm, to which I wrote, seeking access to a few of the hair strands. However, the farm's administration refused to consent to the destruction of those hairs for the purpose of our mtDNA analysis. Their reply also indicated their unshakable conviction that Jesse James and no one but Jesse James had been killed in St. Joseph, Missouri, in the Lafayette Street home of Thomas Howard, alias Jesse James, on April 3, 1882.

That refusal left this project with only one alternative. All other reasonable means to make an identification had been exhausted. An exhumation was the last and only resort.

ON FRIDAY, MARCH 19, 1995, DR. GEORGE STEPHENS AND I WENT TO the grave site of the James family in Mount Olivet Cemetery in Kearney, Missouri. We found the burial site at the west end and on a level plane in the relatively treeless Mount Olivet Cemetery, a grassy place with headstones of differing structure and quality of granite. The grave marker for Jesse James was flush with the ground, with Zerelda, his mother, on one side of him and Zee, his wife, on the other. Zerelda had her own monument, while Zee shared the flat marker with Jesse. Carved into the stone were their birth and death dates, and Jesse's read: "Jesse W. Born Sept 5, 1847, Assassinated April 3, 1882." In the same general area were the graves of his half brother, Archie Samuel, whom the Pinkertons had killed, and his stepfather, Reuben Samuel. Behind Jesse's marker was his white Civil War military monument, attesting to his service to the Confederacy.

The burial plot, lot number 91 in block number 1 of the original section of the cemetery, was in reality only half the usual lot size of twenty feet, with the ten-foot side running east and west, while the twenty-foot side had a north-and-south course. The size of the grave site, as well as the number of persons listed as buried there in the James family, created a logistical problem. We had to ensure that only the grave marked as that of Jesse James would be the focus of our exhumation. An entry into any other grave would be legally and ethically unacceptable. It was going to be tight.

Dr. Stephens performed various geological analyses. He noted that with a slight slope from east to west, drainage was to the west. The grave site was five to six feet below the crest of a hill. He collected soil samples one foot west of the northwest corner of the marker stone for Jesse and Zerelda James and at another point near the southern edge of the cemetery. The soil profiles, back at his laboratory, proved to be identical. It was heavily organic, very clay-rich soil with high moisture retention. The pH reading varied from 5.5 to 7.5. To our distress, augering demonstrated a high water table. The suction from the water-laden subsurface soil was so great that it took the two of us to wrench the auger from the ground. Leaving it standing in place would make it appear as if someone had driven a dagger into the James family's plot. Jesse James was many things, but he was no vampire.

Upon our inquiry, we were relieved to hear that Kearney had had an unusually high amount of rain during the spring of 1995. That meant there should be no unexpected pools of water in the grave pit resulting from large amounts of rain over the long term doing serious damage to the coffin and/or the remains. Hopefully, we could rely on the meteorological reports to that effect.

The records indisputably indicate that the interment of the supposed Jesse James's remains at Mount Olivet Cemetery was on Sunday, June 29, 1902, two years after Zee, his wife, had died. His removal from the farm grave and reinterment at Mount Olivet were now in our investigative sights.

Stella James, wife of Jesse Edwards James (Jesse James's son), and Robertus Love, a newspaper reporter, gave accounts of Jesse's disinterment, both concurring that the transfer from his seven-foot-deep grave was accomplished with much difficulty. The weather was entirely uncooperative that Sunday, causing the exhumation to take four hours. The disinterment was undertaken in the midst of a heavy, soaking rain, making the task one that everyone wished to expedite. The inclement weather also caused the absence of Frank James, who was ailing.

But it wasn't only the weather that conspired against a speedy and problem-free transfer. The coffin, apparently thought to be intact after twenty years in the ground, collapsed. When the metal casket lid containing the viewing glass was lifted out, the bottom fell back into the grave. To put the best spin on it, the scene must have been a dispiriting jumble, with funeral clothing and bones spilling out into the grave pit while drenched and horror-stricken bystanders looked on. The skull fell back into the grave once, and in the attempt to retrieve it, it fell in a second time.

From all reports, Jesse Edwards James was called upon to examine his father's defleshed and disarticulated skull. It is said that he noted the presence of gold fillings in the teeth. These gave some support to its being his father's skull, since Jesse was known to have prided himself on his gold fillings. More important, Jesse Edwards took note of the entrance wound in the back of the skull, giving more substance to his identification of it— that is, if his recollection twenty years after seeing it in death in his home in St. Joseph can be counted on as reliable. Standing at the grave in the pouring rain, Jesse Edwards pronounced the skull as that of his father.

THE FIRST STEP TO BE TAKEN UNDER THE MISSOURI STATUTES governing the exhumation of human remains was to have Dr. Gerald B. Lee, the elected medical examiner of Clay County, forward a request to the prosecuting attorney for Clay County, Michael E. Reardon, to initiate the legal proceedings toward the issuance of a court order for the exhumation of the remains of the supposed Jesse James. Dr. Lee was enthusiastically agreeable to forwarding such a request, which was done on May 26, 1995. To be on the safe side—even though unnecessary under the terms of the relevant Missouri statutes—that letter referenced the fact that many descendants of Jesse James had consented to the request for an exhumation.

After a number of telephone contacts and a personal visit to Prosecuting Attorney Reardon, his office filed a "Motion to Exhume the Body of Jesse James with Suggestions" on June 19, 1995. Nine days later, on June 28, an amended motion, correcting and elaborating on certain details in the original motion, was filed. These filings set the stage for a court hearing in Clay County on the petition to exhume.

The judge and prosecuting attorney quizzed me concerning my contacts with the descendants of Jesse James and their attitudes to the proposed exhumation. I testified that seventeen persons claiming to be Jesse's descendants had consented, and these notarized consents were introduced into evidence. I also pointed out, without mentioning any name, that one relative who refused to sign a consent had "expressed reservations."

One of the consenting claimed descendants was Helen Dillon, wife of the team's handwriting expert, Duayne Dillon, D.Crim. In her case, as in that of the other sworn consenters, I did not deem it necessary to verify their genealogical link to the James family, so long as they stated in good faith that such a relationship existed.

Further, I testified that a Texas attorney, one Waggoner Carr, Esq., had been in touch with me on behalf of his clients, who claimed to be descended from a Jesse James (aka J. Frank Dalton) who had in their opinion survived Bob Ford's shooting in 1882. Mr. Carr had informed me that his clients were wholeheartedly in favor of the exhumation. Apparently, there was some lingering question over the ownership of an as-yet-undiscovered treasure that Jesse James was thought to have buried. These Texas residents were interested in sharing in that buried treasure, if it was located in the grave of their purported ancestor, as well as in being identified as legitimate descendants of the outlaw Jesse James.

In the courtroom, I noticed that Betty Barr and her mother were seated toward the back and I quivered at the possibility that they might have changed their minds from the time of our original meeting. When the judge asked if there were any objections to the exhumation, I watched them with bated breath, but they sat without uttering a word. The judge waited, then waited some more. I felt panic rising in me. Still, no voice was heard in the courtroom. Finally, to cut off any belated objections that the extended silence might bring (something like being at a business meeting where a long silence always results in a continuation of the debate), I nudged the assistant prosecutor, urging him to have the judge note the absence of any objections. He did so, and the requested court order was rendered. At last, the exhumation could commence.

The order was brief, one page in length, and its terms were quite explicit. The exhumation was to be conducted prior to July 31, 1995, with the remains to be placed in the custody of Dr. Lee "or his authorized representative," in order to perform an "autopsy and scientific studies." The costs were to be borne entirely by The George Washington University, Professor James E. Starrs, and the nonprofit corporation Scientific Sleuthing, Inc. The remains were to be reburied on or before October 31, 1995, some ninety days or thereabouts from July 31, 1995, the last date permitted for the exhumation to occur.

Coincidentally, it was Judge Howard whose order provided the requisite authorization to exhume the man who at his death was living under the alias of Tom Howard. Was this an omen auguring for success in our exhumation? As another such augury, I chose July 17, 1995, my grandson Willie's tenth birthday, as the date of the exhumation. Surely good fortune would shine on me with Judge Howard and Willie in my corner. Not until many days later, when the exhumation was a done deed, did I realize that Judge Howard was without statutory authority to issue a court order for the exhumation. Rightfully and legally we should have petitioned a sister court in St. Joseph, where the death occurred. The gods seemed to be smiling on me from all sides. But would my good luck hold?

The exhumation was to begin with the breaking of ground at Mount Olivet Cemetery on July 17, but there were still additional preliminaries to be accomplished. On Sunday, July 16, George Stephens and I mapped the ten-by-twenty-foot grave site so as to ensure that all headstones and other markers would be returned after the exhumation to their original locations. After this survey, we conducted metal-detector runs over the site, and George scanned the graves with a magnetometer. We wanted to ensure that only the grave of the purported Jesse James was disturbed and no other. We also sought to assess the presence of a metallic coffin, as described in a *New York Times* article about the 1902 removal. Dr. Stephens reported that the results of his surveys indicated the presence of numerous small metallic wires in the ground, probably from flowers bundled and left at the site.

We also prepared the site for the next day's exhumation by moving the footstone from Zee James's grave, as well as by loosening other stones that would have to be temporarily dislodged during the exhumation. We hoped that by doing this we could accelerate Monday's exhumation. Little did we know what surprises the next day would bring.

Fortunately, the weather was rain-free for the entire period of our outside work, although it was blazing hot, on all days hitting highs in the nineties. The heat necessitated a tent to shelter the grave pit during the exhumation. We knew from experience that unless we shaded the bones once they were exposed, they would be subject to rapid artifactual alteration from the drying and cracking effect of the sun's sudden and intense heat.

The mob scene at Mount Olivet on the morning of July 17, 1995, was something that punctilious scientists would rather forget, or better, not have to confront and combat. The media, print and broadcast, were massed in numbers that defied an accurate tally. Vehicles with satellite dishes were assembled at a safe and noninterfering distance. Spectators from near and far, and relatives of the real and the not-so-real Jesse James, vied for spots within eye- or earshot of the excavation site. Fortunately, we had cordoned off an area with barrier tape for our work area.

Outside that tape the press were champing at the bit for a regular stream of information. My ten-year-old grandson, Willie, became the object of their attention when they learned that July 17, the day of the start of the exhumation, was his birthday. A team from the *Good Morning America* program miked Willie and me for a live interview. But Willie froze at this, his first, TV opportunity. Later he explained that his earphone was not working so he could not hear their questions. Indeed, the background noise was enough to deafen anyone.

For security purposes, I approved the presence of uniformed guards from the Pinkerton Agency. Some of my team members saw this as an ironic twist, as I intended it to be. During Jesse's lifetime, these were the men hired to find him and gun him down. So I decided "Let's make amends" by giving them a role in protecting the grave site and in crowd control. Representatives from organizations representing Confederates were, needless to say, not pleased by the Pinkerton detectives' presence.

The first action we undertook was to remove all monuments and other obstructions to the east of the ground-level headstone, leaving just the James headstone in place. We then brought a backhoe into action, which had a shovel no wider than that of the usual grave.

The grave was backhoed to a depth of first one and then two feet. At each measured one-foot depth, I would probe its east end with a four-foot-long metal rod in hopes of striking an obstruction, either a metal or a wood casket. (The east end would be the foot end in the Christian burial tradition. Doing damage to the remains at that end would not be catastrophic.) On the contrary, the rod gave no indication that an intact coffin existed at the site at all. When we had gotten down five feet into the ground, I realized that my probing for a solid container was fruitless. I saw wood shards on either side of the pit, telling me that it was time for the exhumation to become an excavation and that bamboo sticks, trowels, and brushes were the implements of choice for that purpose.

Clearly, there would be no coffin to lift out simply and swiftly, and I realized a dramatically new tack was in order. What was once planned to be a quick and easy exhumation of a metallic coffin would now become the more time-consuming and rigorous process of an archaeological dig. However, time was pressing. The custodian of the Mount Olive Cemetery, Mr. Darrell Cave, had at the last moment advised us that a burial was set to occur at the cemetery on Wednesday afternoon, the nineteenth of July. Consequently, he required that we complete our undertaking at the Jesse James grave site prior to noon on that day. We had less than two days to perform a demanding task that would require the greatest of care.

With this time limit confronting us, we had to dispense with some of the more exacting and time-intensive features of an archaeological exploration. For example, we had to forgo the idea of measuring and marking the exact location of all bones and artifacts that we might find in the grave pit. Only accurate but rough estimates of distances from point to point could be taken. This was an unexpected and dispiriting turn of events.

THE FIRST ITEMS WORTHY OF MENTION FOUND IN THE PIT WERE shards of decaying wood, presumptively from a coffin. From that discovery it became evident that the contents of the grave, such as they were, would be revealed piecemeal. And such was in fact the case.

Working from the westernmost part of the grave pit (nearest the headstone), Dan Kysar, an experienced archaeologist and former student of Dr. Michael Finnegan (the man who had been part of the 1978 exhumation at the James farm), ploddingly and conscientiously exposed several artifacts. Then he turned up badly decayed skeletal remains. My daughter-in-law, Traci Starrs, a forensic scientist in her own right, was carefully recording each artifact that he turned up. Uncomplaining and conscientious, she sat right by the grave pit in the hot sun, ensuring that the process was done correctly and that all retrieved items were documented. It was a good thing for us that she was so observant. She would soon make a surprising and important discovery.

As we watched, standing about the pit, a glass viewing plate, fractured into many pieces but still in its original orientation, was lifted out bit by bit. Then uniquely ornamented swing bale handles, with pieces of wood still positioned in them, were exposed, bagged, and documented on chain-of-custody forms. To our relief, the first day ended with the vague outline of skeletal remains revealed, but we left them in place, in the Pinkerton Agency's care, until our next day's labors. The local sheriff and his men shared the night-patrol duties. The grave site was to be doubly protected.

The second and third days of the exhumation followed a plan that involved a strict division of responsibilities. We had to act expeditiously. While some team members bagged and logged artifacts and bones at the grave site, another group, with Dr. Jay Dix at the helm, was actively engrossed in X-raying and examining everything that had been bagged and shipped to the Kansas City Regional Crime Laboratory, the hub of our activities away from the cemetery.

In the course of the radiographic work on a mass of debris under the sure and competent command of radiographer Michael Calhoun, a small square object was detected. Since Jim Kendrick, our team's photographer, made the find on the X-ray, he was put to the task of locating the item. It turned out to a blackish square stone, five millimeters square, that appeared to be the end of a tie tack.

Thinking that my grandson, Willie, deserved a lesson from me in geology, I asked him what he thought the stone might be.

Without a moment's hesitation, he replied: "Obsidian, Grandpa."

Floored by the accuracy of this ten-year-old's opinion I asked him, in something of an attempt to put him in his place, "And what is obsidian?"

Again, in less than a metronome's slow beat, he answered: "Volcanic rock, Grandpa."

I carried the conversation no further, realizing that his phenomenal accuracy deserved no rejoinder except for the gentle pressure of the arm I put around his shoulders.

At the same time that the X-rays at the laboratory were creating a revelatory stir, Dan Kysar made a stunning find at the grave. Amid fragments of the remains' right ribs, he unearthed a spent bullet. This item was immediately presented to the team's firearms expert, John Cayton, who declared it to be a .36-caliber lead ball. He posited that the ball had been fired either from an 1851 Navy Colt, a weapon commonly possessed by Union officers during the Civil War, or from some other .36-caliber handgun with lever action. This information added weight to the mounting evidence that the exhumed remains were those of Jesse James, who was thought to have carried two bullets from antemortem wounds to his grave. Why not? At that time, lead poisoning was less of a fatal danger than death from surgical intervention.

If nothing else, this find proved that the remains that were under investigation in the grave were most assuredly those of a man, not of either Jesse's mother or his wife. In addition, the bullet's being found among the badly decayed right rib fragments served circumstantially to connect the remains to Jesse James, who was known during his lifetime to have carried at least one bullet in his right chest. He had been shot there and severely injured in 1865 while attempting to surrender to the Union forces at Lexington, Missouri.

At this point, we turned to the detailed anthropological analysis of the various parts of the skeletal remains we had recovered. Dr. Michael Finnegan reported through sex, age, stature, and racial typing that the exhumed bones fit the known profile of Jesse James. The skull, reconstructed with deft care and Duco cement by Dan Kysar, had shown clear evidence of one, and only one, entrance wound, located behind the right ear. No exit wounds were discovered in the skull, either in the frontal bone (the forehead), where some thought one would be found, or elsewhere. At a later time, a qualified firearms examiner, Jay Mason, performed a two-part chemical test on an occipital bone fragment that possessed a whitish stain indicative of possible lead oxidation. The test (sodium rhodizonate) demonstrated conclusively the presence of lead from the passage of a lead bullet, a feature known as lead wipe.

A computer animation of the shooting of Jesse James by Bob Ford was later designed by Laura Denk of Engineering Animation, Inc. This animation graphically and compellingly illustrated the version of one fatal bullet with no exit and a laceration to the left forehead caused by Jesse James's falling to the hardwood floor after being shot. We concluded that the bloody abrasion on Jesse's forehead, by some thought to be an exit wound, was likely the result of hitting his head when he fell.

In the end, the James farm relented and allowed us to take the hair from the James farm museum and test it. Dr. Walter Rowe of the Forensic Sciences Department of The George Washington University reported that the hair showed unmistakable signs of having been buried and of having been dyed, apparently in an effort to disguise the owner's identity.

Then, since some persons had intimated that the pain from Jesse's many serious wounds had led him to become addicted to drugs, Dr. Bruce Goldberger, an esteemed toxicologist with the University of Florida Medical School, subjected the hair to toxicological analysis. He reported that it revealed no traces of opium, cocaine, or other derivatives. The absence of such drugs or their metabolites proves either that Jesse James was not using these drugs within the three-month period prior to his death or that the presence of these substances had been eliminated by degradation or otherwise over the years.

During Dan Kysar's painstaking work in the grave pit, he managed to find sixteen teeth, each of which was no larger than half the size of my thumbnail. Dr. John McDowell, an odontologist with the University of Colorado Medical School and soon to be president of the American Academy of Forensic Sciences, analyzed them—with surprising results. His report added new biographical details to those already known about Jesse James. The teeth showed nicotine staining and the corrosive influence of regular tobacco chewing. Jesse's previous known involvement with this vice had been limited to a report of his having killed a companion in a dispute over chewing tobacco.

Dr. McDowell also postulated that the reason Jesse had not posed for any photographs with a smile on his face might have been his lack of an upper right central incisor, which would have produced a rather unsightly gap-toothed smile. His examination of maxillary (upper-jaw) fragments from the exhumed grave led him to conclude that at some point during his adult lifetime Jesse had lost the upper incisor.

The numerous gold-foil fillings in the teeth—the very fillings that Jesse Edwards had noted during his father's 1902 exhumation—signified that Jesse James had consulted dentists who were progressive in their techniques and that Jesse was willing to pay the price for such advanced dental care. These fillings did correspond to what was known about Jesse James during his life, except that the gold restorations were found in many more teeth than we had anticipated. Was it possible that Jesse James took his ill-gotten gains to the grave with him in the form of gold-foil tooth restorations?

It became to some extent a bone of contention among Jesse's family members as to who would get the gold from the teeth, but in the end, all the teeth went back into the ground, the gold with them, except for those teeth destroyed in our DNA analysis.

Every precaution was observed to ensure the reliability of the team's ultimate conclusions on the identity question. To that end, Dr. Duayne Dillon, a handwriting expert of note, was called upon to examine and compare the existing handwriting specimens of Jesse James with those of J. Frank Dalton. The analysis proved that Dalton and James had distinctly different handwriting characteristics. Consequently, this assessment added support to the assertion that Dalton was not Jesse James.

All of this provided mounting evidence that we did indeed have the remains of Jesse James. We had only to look at the mtDNA analysis, which would tip the scales conclusively and irrevocably.

The report by Dr. Mark Stoneking and his doctoral student, Anne C. Stone (now Dr. Stone), then of Pennsylvania State University on the results of the mtDNA testing was the last and most compelling item on the team's agenda. Stoneking and Stone had tried getting DNA from the bones from the grave, but their decayed condition made that impossible. As a last resort, they turned to the teeth. Teeth have protective enamel coating, and inside that coating the DNA is housed. So the enamel protects the DNA, and we had sixteen teeth available to test. After one tooth after another tested out unsuccessfully, Stoneking and Stone finally announced a match between two molars and the mtDNA of attorney Bob Jackson and Mark Nichols, the two known descendants of James's sister, Susan Lavenia James. The sequence of base pairs was so singular that it was said to be the first time it had been encountered in the entire mtDNA database for the Northern European population.

At last, with DNA, we had untied the Gordian knot! The newspaper had said, "Jesse! By Jehovah!" Now it was "Jesse! By DNA."

But that still wasn't all: I have one more unusual event yet to report. With any exhumation, there is the potential for what I call the terra incognita, or the unknown ingredient. We can never fully predict what we might find. Dr. Finnegan had highlighted a perverse and mysterious feature of the 1902 reburial of Jesse James that came to public notice only as a consequence of our exhumation.

As Dan Kysar was taking the bones out of the grave, one by one, Traci Starrs described and catalogued them. At one point, she stopped the operation in its tracks by remarking, "Something's wrong." She pointed to the indentations in the soil of the grave pit, still clearly seen, where the bones had been positioned. Having an anatomical chart with her and standing at the foot of the grave, she informed Dan Kysar that all the right-side bones were on the right side of the skeleton and all the left-side bones were on its left. Consequently, Jesse James had been reburied in 1902 in a prone and anatomically correct position. They looked at each other in astonishment. Both spoke at the same time and in agreement. Jesse James had been buried facedown!

The reason for this unorthodox burial is unknown, but we speculated that the person who arranged the bones in 1902 in this peculiar manner might have done so to inflict a final indignity upon Jesse. Rather than looking to the heavens whence his savior would arrive at the Second Coming, as the Bible has it, Jesse was placed facing the proverbial fires of hell. There's an enigma to engross and to perplex future Jesse James researchers.

With the resolution of the mystery of the identity of the remains, a new mystery has now emerged to cause consternation and debate among future generations of historians of the West.

Jesse was buried as New Mexico's legendary train robber Thomas "Black Jack" Ketchum had wished. It is reported that Ketchum declared his desire to be buried facedown after his execution in April 1901, so that, he is reputed to have said, the world could kiss his ass. His hanging, however, was not followed by the granting of his wish. Jesse got the benefit, such as it was, of the facedown burial Ketchum desired even without being known to have sought it. Another feather in Jesse's cap of legends?

Finally, we transported the remains back to The George Washington University for analysis in our lab and to arrange for anthropologist Diane France, Ph.D., to create a cast of Jesse's skull. I decided to carry the boxed skull myself on the plane, and to steady it, I placed four bottles of Guinness (my favorite beverage) around it. The container was rather large, too large to fit in the overhead space or under the seat in front of me. A flight attendant stopped next to me and told me that I would need to place the box either overhead or under the seat in front of me. She volunteered to put it with check-in baggage, but I vetoed that proposal.

"What's in it?" she asked.

I looked at her, wondering whether to tell the truth or not. She persisted, so I sheepishly told her it was the skull of Jesse James. That seemed to surprise her. She wanted me to open the box for confirmation, and when I did, she spotted the four bottles of Guinness positioned around Jesse's reconstructed skull.

"What are *they* for?" she asked.

I promptly replied, "Just in case he gets thirsty."

THE SCIENTIFIC INVESTIGATION INTO THE DEATH OF JESSE JAMES was a total, a smashing, and an incontestable success. The scientific team accomplished every objective set before it and even discovered items of relevant information that were significant to the identification but wholly unexpected.

The results of this exhumation conclusively proved the inestimable value of exhumations in clarifying and rectifying the historical record. No longer could those skeptical of such scientific activities protest that they have no merit. The dissenters were silenced and solace was given to the authentic descendants of Jesse James. In addition, a new and exciting chapter had been added to the history of the American West.

We came as close as science has ever come to apodictic certainty on the question of the identity of these remains. The uniqueness of Jesse James's mtDNA gives further proof that he deserves not only the special place he has held in the history of the American West but also his newfound distinction in the annals of mtDNA testing.

Jesse James died on April 3, 1882, in St. Joseph, Missouri, when he was shot in the head by a single bullet that did not exit his skull. In light of these scientific findings, the claims of those who say that someone else died in Jesse's place and that Jesse lived on to father additional children are worse than nonsense. They are ludicrous in the extreme.

In escaping every trap set to capture him, Jesse had played the role of the Houdini of western outlaws. Whereas he may have out-Pinkertonned the intrepid Pinkertons during his lifetime, he could not outfox science after his death.

5. MARY A. SULLIVAN

She Was Just Nineteen

Tread lightly, she is near
Under the snow,
Speak gently, she can hear
The daisies grow.

All her bright golden hair
Tarnished with rust,
She that was young and fair
Fallen to dust.

OSCAR WILDE, "REQUIESCAT"

By now I had been inspired to examine other historical deaths heavily laden with mystery and controversy, such as the untimely and most tragic demise of famed explorer Meriwether Lewis, said by some historians to have killed himself in 1809. I believed there was good and substantial reason to argue that he was more probably murdered and that modern forensic science could unriddle that mystery, but it would be like untying the Gordian knot to secure the permission of the National Park Service for me to exhume his remains. I petitioned for its consent and unflaggingly continued to press my petition for an exhumation, but in 1998 another historical controversy came knocking at my door, or rather, as it actually was, ringing on my office telephone to distract me from my earnest efforts on behalf of 160 relatives of Meriwether Lewis and the governors of Virginia, Tennessee, and Missouri.

Relatives of a nineteen-year-old woman named Mary A. Sullivan, said to have been the final victim on January 4, 1964, of serial killer Albert

DeSalvo, dubbed the Boston Strangler, believed that despite his arrest and confession of having murdered her, for which murder he had never been tried, her case was not yet closed. In the minds and hearts of Mary's family, DeSalvo had not been the right target suspect. They firmly believed the police had ignored clues that might have led to the real killer. It was, to them, possible, even thirty-four years later, that her killer was still at large.

So with the well-publicized feats of DNA ringing loud and clear, they wanted to put DNA to work in their cause. But the Boston and Massachusetts authorities had no willingness to help. As in the Huey Long investigation by me, the law-enforcement establishment had written in unerasable stone that Albert DeSalvo was her killer. Better try to move a mountain than the Massachusetts authorities.

It was Casey Sherman, Mary Sullivan's nephew, who brought the death of his aunt Mary to my attention. He had seen me on a Discovery Channel documentary about the Jesse James exhumation and had decided to telephone me. He provided me with a précis on Mary's death, not realizing that I had long ago read Gerald Frank's book on the Boston Strangler. He was passionate to overflowing with a desire to get to the truth of who killed his aunt Mary.

After an extensive review of the historical record, I determined that it was vital to exhume Mary's remains. Little did I then realize what finds I would uncover and how those findings would explode the received wisdom on the guilt of Albert DeSalvo.

To understand why this case is considered of continuing controversial interest and why we encountered such stiffened resistance to reopening it, the events in Boston some forty years ago that led up to Mary's death must be examined.

THE MOST COMPENDIOUS INFORMATION ON THE SERIAL RAPES and murders by the Boston Strangler comes from a number of books devoted to the killings, including *The Boston Stranglers* by Susan Kelly, *The Boston Strangler* by Gerald Frank, and *Confessions of the Boston Strangler* by George William Rae, with Albert DeSalvo credited as coauthor, which made claim to having transcribed much of the lengthy audiotaped confessions obtained from DeSalvo by the police.

As the tale unfolds, between June 14, 1962, and January 4, 1964, eleven women in the Boston area were victims of either a single serial killer or possibly several killers. The murders were generally believed to be linked to one offender, who became known variously as the Phantom Fiend or Boston Strangler. Within ten weeks in 1962, six women were killed—the first four within twenty-seven days. Then two were killed in the same month. All of these victims were elderly.

The second wave of murders began in December 1963, with younger women the victims. Then, in the spring of 1963, two more elderly women were murdered before the killer turned again to younger women.

Each woman was attacked in her apartment (except for one in a hotel room, who might not have been related to the others), was sexually molested, and was strangled with articles of clothing (except one who was stabbed multiple times). With no sign of forced entry, the women apparently knew their assailant or, at least, had voluntarily allowed him into their homes. It seemed unlikely, especially with the extensive news coverage, that they had left their doors unlocked or opened their doors to complete strangers. Most of these women led quiet, modest lives and were single.

Of the eleven confirmed stranglings (DeSalvo's confession added two more), six of the victims were between the ages of fifty-five and seventy-five. The remaining five victims ranged in age from nineteen to twenty-three. Only one was a non-Caucasian, twenty-year-old Sophie Clark.

The first known victim was Anna Slesers, fifty-five, a seamstress making $60 a week. She lived in a third-floor apartment on Gainsborough Street in Boston's Back Bay area, into which she had moved two weeks earlier. On the evening of June 14, 1962, she had finished dinner and awaited her son to pick her up for church. But just before seven o'clock in the evening, her son found his mother dead on the floor of her apartment, with the cord from her bathrobe wrapped tightly around her neck. Her legs had been positioned for shock effect.

Strangely, the belt from her robe had been knotted tightly about her neck, its ends turned up to look like a bow tied under her chin. There was blood on her right ear and a laceration on the back of her skull. Her neck was scratched and abraded, and there was a contusion on her chin. She had been sexually assaulted with a hard object, possibly a bottle.

Ms. Slesers's apartment appeared to have been ransacked. Her purse was open and its contents strewn on the floor. Trash lay around a kitchen wastebasket, and drawers had been left open in the bedroom dresser, their contents rumpled. A case of color slides had been carefully placed on the bedroom floor. Yet a gold watch and other pieces of jewelry were left untouched. The police assumed that the crime had begun as a burglary and had turned into a sexual assault and murder. There were no immediate suspects.

Two weeks later, on June 30, around 5:00 P.M., sixty-eight-year-old Nina Nichols was murdered in her fourth-floor apartment at 1940 Commonwealth Avenue in the Brighton area of Boston. She had interrupted a call from a friend to answer the door and had not called back. The friend contacted a janitor and asked him to check.

He found the woman on the bedroom floor, bleeding from both ears, with her legs spread wide, her pink housecoat open, and her slip pulled up to her waist. On her feet were blue tennis shoes. Around her neck were two of her own nylon stockings, with the ends tied into a bow. She had been sexually assaulted. Blood was found in her vagina and there was a small abrasion on her face. Her killer had also bitten her. It was later determined that she had been penetrated with a wine bottle.

As with Ms. Slesers, there was no sign of forced entry into her apartment, but the place looked as if it had been burglarized: Every drawer had been pulled open, and items from them lay scattered on the floor. Yet those valuables that one would expect a burglar to take had been left behind. In fact, it was soon determined that nothing of value had been taken. Yet the intruder had gone through Ms. Nichols's address book and mail. The retired physiotherapist had led a modest life. Widowed for two decades, she had no male friends. Once again, there were no obvious suspects.

On the same day, fifteen miles north of Boston in the suburb of Lynn, Massachusetts, Helen Blake, sixty-five, met a similar death, but she was not found until July 2. She had been strangled with two of her stockings. She was found lying facedown on her bed with her legs pulled apart, dried blood in both ears, and her face turned to the left. Her brassiere had been looped around her neck over the stockings and tied in a bow. She had been sexually assaulted, but there were no traces of semen.

Once again, the apartment had been ransacked, and this time it appeared as though two rings had been pulled from her fingers and taken. They were later determined to have diamond settings. The killer had also tried unsuccessfully to open a metal strongbox and a footlocker.

The next fatality occurred on July 11. This murder is included on some lists among those linked to the Strangler, but not on others. Margaret Davis, sixty, was found in a hotel room. She was manually strangled and left in Room 7 of the Hotel Roosevelt on lower Washington Street, Boston. A hotel maid found her lying naked on the bed. A man had checked in with her the night before, but the address he gave on the register was phony. Aside from strangulation, there were no other glaring similarities to the previous crimes.

On August 21 at 7:47 in the evening, the police were called to another murder scene—that of seventy-five-year-old Ida Irga, a shy and retiring widow. Found by her younger brother, she had been killed two days earlier in her fifth-floor apartment at 7 Grove Avenue in Boston's West End. As with the other deaths that were gradually being linked by a police investigation, there was no sign of forced entry.

When discovered, she was lying on her back on the living room floor, wearing a torn nightdress. A white pillowcase was knotted tightly around her neck. Her legs were spread open, with her feet propped on chairs. A bed pillow had been placed beneath her. Her head was covered in blood.

The bedroom light was on, and on the floor police found a sheet and blanket, a pair of women's underpants, two brown hair combs, and a quantity of dried blood. There was a trail of blood leading from the bedroom to the living room, indicating she had been dragged from one room to the other. She had clearly been positioned to give those who would find her a shock they would never forget.

The police noted slight injuries to her genitals, but there were no sperm in her vagina. She had died of manual strangulation and strangulation by ligature. She was called the fifth victim. Again, a check on her acquaintances revealed no potential suspects.

Then, within twenty-four hours, a sixty-seven-year-old nurse named

Jane Sullivan was killed in her apartment at 435 Columbia Road in Dorchester, across town. Yet she was not found for ten days. She had just moved to these quarters during the previous month.

She was on her knees in her bathtub, with her feet up over the back of the tub and her head underneath the faucet. Her face was partly submerged in six inches of water. Her underpants were pulled down, while her housecoat covered the upper part of her body. Her bra was on the bathroom floor. She, too, had been strangled with two nylon stockings, probably in the kitchen, bedroom, or hall where blood was found on the floors. Her hyoid bone was fractured. There were bloodstains on a broom handle and matted blood on the right side of her scalp. There was no sign of forcible entry, nor was the apartment ransacked, even though Ms. Sullivan's purse was found open on the sofa.

An unknown killer was terrorizing Boston. Single older women no longer felt safe in their homes, because it seemed as if this phantom could get in and out of their apartments at will.

Then, after a three-month break in the killings, the homicides shifted in a puzzling manner, enough to turn a serial killer's profiler to jelly.

ON DECEMBER 5, 1962, SOPHIE CLARK, TWENTY, AN ATTRACTIVE African-American student at the Carnegie Institute of Medical Technology, was found murdered in her apartment. The apartment, which she shared with two roommates, was at 315 Huntington Avenue in Boston's Back Bay area, a couple of blocks from Anna Slesers's apartment.

Sophie lay in the living room, partially clad, wearing a print housecoat, a black garter belt, black stockings, and black tie shoes, with her legs spread wide apart. She had been strangled with a nylon stocking that had been knotted and tied tightly around her neck. Her half-slip had also been tied around her neck, and a gag placed over her mouth. There was evidence of sexual assault, with semen being found on the rug near her body. Sophie wore a sanitary belt. Nearby were a ripped bra, bloodstained underpants, and a sanitary napkin. She had suffered no genital injuries.

Again, there was no sign of forcible entry, and Sophie was known to be cautious and security-conscious. Even so, her killer had somehow convinced her to let him in. He had struggled with her until she was dead and then had rummaged through drawers and gone through her collection of classical records.

Sophie had a boyfriend in New Jersey but did not date anyone in the Boston area, so there were no fruitful leads.

There were some differences between this and the earlier Strangler murders. Sophie was black, twenty, and did not live alone. Also, for the first time, there was evidence of semen at the scene of the crime.

When the police questioned Sophie's neighbors, they discovered that a man wearing a dark jacket and green trousers had knocked on doors, making a false statement that the superintendent had sent him. He then made inappropriate remarks to women about their bodies. One resident described him as being between twenty-five and thirty years old, of average height, and with honey-colored hair.

There was some question as to whether this murder fit with the others, in particular because there were other possible motives. For eighteen months prior to Ms. Clark's murder, classmates of hers had been receiving threatening messages. This was the early 1960s, and an area of Boston that involved a great deal of civil unrest and racial attacks. Someone had made clear to Sophie's roommates at another apartment, one of whom was black and one white, his belief that it was not appropriate for black women and white women to be rooming together. The women had moved together to Huntington Avenue, but the threatening phone calls had continued. This murder could have been racially motivated.

Three weeks later, on Monday, December 31, 1962, Patricia Bissette, twenty-three, single, and a secretary for a Boston engineering firm, was discovered murdered when her boss arrived to pick her up for work. Her apartment building at 515 Park Drive was in the Back Bay area in which Anna Slesers and Sophie Clark had lived.

Her discoverers found her faceup in bed with the covers drawn up to her chin. Underneath the covers, she lay with three stockings and a blouse knotted tightly around her neck. There was evidence of recent sexual intercourse, and during the autopsy she was found to be one month pregnant. There had been some damage to her rectum. As with the other murders, the killer had also gone through her personal items.

Bissette had lived alone. Skilled in math, science, and physics, she had participated in numerous social organizations and was a regular churchgoer. At 4:30 on Sunday afternoon, a waitress had noticed her with a dark-skinned man who spoke with an accent. After that, no one reported having seen her. It eventually came out that she had been involved in an affair with her married boss, Jules Rothman, and was also dating another man. His remarks to the police after the murder were suspicious but were not sufficient reason to hold him. Being meretricious is not necessarily being a murderer; otherwise, think of the vast societal consequences.

Four months went by without another murder that fit this pattern in any way, although on March 6, 1963, twenty-five miles north of Boston in Lawrence, sixty-eight-year-old Mary Brown was found beaten to death in the living room of her apartment at 319 Park Street, where she lived alone. She had also been strangled and raped, but her death was caused by blows to her head. She was nude and her girdle was pulled down to her left foot. She still wore her rubber overshoes and stockings. Pulled over her head were a black dress and other articles of clothing. Her throat was bruised, and a kitchen utensil was stuck deep into her left breast.

Two months later, another murder occurred in Boston. On Wednesday, May 8, 1963, Beverly Samans, a pretty twenty-six-year-old graduate student, missed choir practice at the Second Unitarian Church in Back Bay. A friend checked her apartment at 7:00 P.M. and opened it with the key she had given him. He found her lying nude on the sofa bed directly in front of him, with her legs spread apart. Her hands, each wrist individually tied, were bound behind her with one of her scarves. Two nylon stockings and a white scarf were knotted loosely around her neck. A cloth had been placed over her mouth, covering another cloth that was stuffed into it. There were no ligature marks on her neck.

While it appeared that Beverly Samans had been strangled, she had, in fact, been killed by four stab wounds to her throat, two to the left and two to the right. In all, she had sustained seventeen stab wounds, the majority of which were to her left breast in what appeared to be a bull's-eye pattern.

The ligature around her neck was not tied tightly enough to have strangled her. A bloody knife turned up in her kitchen sink. She had not been raped, nor was any semen present in or on her body. It was estimated that she had been dead approximately forty-eight to seventy-two hours; she had probably been killed between late Sunday evening and Monday morning.

Although the police had had no suspects for the other killings, they did have in their custody Daniel Pennachio, who had been arrested for lewd and lascivious behavior occurring just shortly after the Samans murder. Pennachio confessed to the Samans murder and gave details to the police that had not been revealed in the media. The police decided that Pennachio's confession was untruthful, and he was ultimately released to die in a diving accident.

Another quiet period of approximately four months ensued during the summer of 1963, but on September 8, the pattern shifted again. At 224 Lafayette Street in Salem, Evelyn Corbin, a pretty fifty-eight-year-old divorcée who looked a decade younger, was found murdered in the bedroom of her first-floor apartment. Her neighbor, Flora Manchester, had last seen Evelyn at 10:30 that morning. Shortly after 1:00, she used the key she had to open the door and found Evelyn dead.

Strangled with two of her nylon stockings, she lay across the bed faceup, her left leg hanging to the floor. The upper left corner of the bedspread covered her trunk. She was dressed in a housecoat, nightgown, and ankle socks, and three buttons were missing from the robe. Around the bed on the floor were lipstick-marked tissues that had traces of semen. Semen was found in her mouth, but not in her vagina. There was blood in her ears and a bloodstain on the bedcover beneath her head. Her right hand and forearm were twisted beneath her body. A stocking was tied around her left ankle, and two around her neck. On the floor lay a pair of women's underpants, blood- and lipstick-stained above the crotch.

Her locked apartment had been searched, but apparently nothing was stolen. A tray of jewelry had been put on the floor, and her purse had been emptied onto the sofa. Outside her window on the fire escape was a fresh doughnut. It was later determined that no one from the building had thrown it there. The police had no leads in this murder.

A forensic scientist in today's high-technology world of accredited crime laboratories and evidence technicians with portable crime-scene equipment of all sorts would be at a loss to explain the total failure of the police to connect these homicides either to one man or to multiple murderers.

The police cannot be faulted for not summoning DNA profiling to their aid, for it would take decades for DNA to be recognized as an identifying tool. But the knots used to tie the ligatures the Strangler was using could have been some evidence of one man or more than one. Were the knots distinctive? Could they suggest whether the killer was right- or left-handed in all or only some of the ligatures?

And where were the fingerprints, the hair foreign to the scene, the shoe outsole prints on the floor and carpets, et cetera? As to fingerprints, it was improbable that the Strangler investigators would be able to visualize the Strangler's latent prints on the bodies of his victims, alternate light and cyanoacrylic ester (superglue) fuming being still years away from recognition as a useful tool in forensics.

Other forensic disciplines still in their infancy in the early to mid-1960s include odontology (the study of dentition) in evaluating bite marks, and criminal profiling, which had already proved its quackery in the search for New York City's "Mad Bomber" (George Metesky) and more recently has had its reliability upended in the tracking of the Washington-area snipers Muhammad and Malvo.

Certainly, as Sherlock Holmes instructed John Watson, crimes are not committed by flying creatures who leave no traces and thereby act in disregard of Edmund Locard's exchange principle. That principle, a staple in forensic science, almost too obvious to require explanation, teaches that when any two objects come into contact, there is a reciprocal exchange of trace evidence from one to the other. Consequently, when a criminal leaves the scene of the crime some evidence of the presence of that person is taken with him, but the criminal also deposits something at the crime scene traceable to him.

Were the Boston police oblivious to Locard's exchange principle, or were eyewitnesses and confessions more the order of their day than hard incriminating or exonerating physical evidence?

Two and a half months went by. Then, on November 24, as the Boston area was grieving the loss of President John F. Kennedy, who had just been assassinated, Joann Graff was found raped and murdered in her ransacked Lawrence apartment.

The religious twenty-three-year-old University of Chicago graduate student and industrial designer was found in the bedroom by her landlord and a Lawrence police officer, there at the urging of her worried friends. She lay diagonally across her bed, her right leg dangling over the side. Her arms were crooked and her right hand, lying on her midriff, was curled into a fist.

She wore only an open blouse. Two brown nylon stockings and a pair of tights had been tied in an elaborate bow—a square knot with an extra loop—around her neck. One stocking had a granny knot and the other a surgeon's knot. Two contusions below her right nipple and two abrasions above and to the left turned out to be teeth marks. The outside of Graff's vagina was bloody and lacerated, and the autopsy showed that she had been raped. Her right thigh had a contusion.

Beneath her body on the bloodstained spread was a torn and bloody bra. Her slacks and underpants had been dropped to the floor, and her eyeglasses lay beside her head.

At 3:25 P.M., the student who lived above her had heard footsteps in the hall near his apartment. His wife had been concerned that someone was sneaking around, causing him to go to the door to listen. When he heard a knock on the door of the apartment opposite his, the student opened his door to find a man of about twenty-seven with pomaded hair, dressed in dark green slacks and a dark shirt and jacket.

"Does Joann Graff live here?" he had asked.

The student told him that Joann lived on the floor below the apartment at which he was knocking. Moments later, he heard the door open and shut on the floor beneath him and assumed that Joann had let the man in. Ten minutes later, someone telephoned Joann, but there was no answer.

And then it was Mary Sullivan's turn, the person of most immediate and overriding concern to my investigation. A nineteen-year-old single woman, she was found murdered at 6:20 P.M. on January 4 by a female roommate in their upper-floor lodgings at busy, mixed-residential-commercial 44A Charles Street in Boston.

According to the reports, Mary was left in a sitting position on the bed, her back against the headboard, her buttocks on a pillow. Her bra had been loosened to expose her breasts, which had been mauled. Over her shoulders she wore an open blouse. Thick liquid that looked like semen was dripping from her mouth onto her right breast. Like the other victims, she had been strangled—first with a dark stocking, then, over the stocking, with a pink silk scarf tied with a huge bow under her chin. Over that was another pink and white flowered scarf. Her knees were flexed, her thighs spread apart. A broomstick handle had been rammed more than three inches into her vagina. Quite grotesquely, a bright "Happy New Year's" card had been propped up against her left foot.

Mary A. Sullivan was the last of the victims in this series of Strangler killings, and quite likely the last of all the Strangler killings.

AS THE POLICE WERE STRUGGLING TO LINK THE MURDERS, PO-lice commissioner Edmund McNamara had already issued a warning to women in the Boston area to lock their doors and be wary of strangers (deadbolt locks were being sold with the frequency of cartons of milk before a paralyzing snowfall). He canceled police vacations and transferred all detectives to Homicide. An investigation began of known sex offenders in the area and of violent former mental patients.

Some 2,350 police would eventually become involved, interviewing a total of around 36,000 people, with suspects into the thousands. Hundreds were fingerprinted and forty given lie-detector tests. Six flunked. Thirty officers patrolled Boston's Back Bay, and every known sex offender was tracked down. Patient leaves from mental institutions were compared against the times of the murders. Seemingly good suspects turned up but for various reasons were dismissed.

A couple of weeks after the murder of Mary Sullivan, Massachusetts Attorney General Edward Brooke took over. At that time, Brooke had the distinction of being the only African-American attorney general in the country. He put together a "Strangler Bureau."

The case spanned five police jurisdictions, with the Bureau coordinating all five locales. Brooke believed that this special task force would mollify the newspapers, especially two female reporters for the *Record-American,* Jean Cole and Loretta McLaughlin, who had made a crusade out of exposing the Boston Police Department's miscues.

To head up this task force, formally called the Special Division of Crime Research and Detection, Brooke chose Assistant Attorney General John S. Bottomly. His team consisted of the Boston Police Department's Detective Phillip DiNatale and Special Officer James Mellon; Metropolitan Police Officer Stephen Delaney; and State Police Detective Lieutenant Andrew Tuney. Dr. Donald Kenefick headed up a medical-psychiatric advisory committee with several well-known experts in forensic medicine.

Two months later, Governor Peabody offered a $10,000 reward to any person furnishing information leading to the arrest and conviction of the person who had committed the murders.

The Strangler Bureau's imposing task was to collect, organize, and assimilate more than thirty-seven thousand documents. A Concord, Massachusetts, company donated a computer to help keep track of it all—reputed to be the first time a computer was brought to bear in a criminal investigation.

Even as they organized all of this material, they had made an arrest in another series of crimes and did not yet realize that they had caught the man who would eventually confess.

On November 5, 1964, based on a police sketch, Albert DeSalvo was arrested for entering women's apartments and raping them. He was known as the "Green Man" because he wore green work clothes. His modus operandi was to force his way in and order the victim at knifepoint to undress. Before leaving, he often apologized.

DeSalvo turned out to be the man police had once arrested in association with the exploits of a sexual deviant known as the "Measuring Man." A few years before the strangling murders began, a series of sex offenses occurred in the Cambridge area of Boston. A charming man in his late twenties would knock at the door of an apartment and, if a young woman answered, would say he was from a modeling agency. When he had them interested, he would take their measurements and often end up in bed with them. Several women reported him, and on March 17, 1961, Cambridge police caught a man breaking into a house who confessed to being the notorious Measuring Man.

Albert DeSalvo lived in Malden, Massachusetts, with his German-born wife and two small children. He worked as a press operator in a rubber factory. The judge, sympathetic because DeSalvo was a breadwinner, had reduced his sentence to eighteen months, and he was released in April 1962, two months before the Strangler first struck.

Soon his background came to incriminate him. Albert DeSalvo was born in Chelsea, Massachusetts, on September 3, 1931. His parents, Frank and Charlotte, had five other children. He reported that his father had been a violently abusive man who regularly beat his wife and children. As a boy, Albert had been arrested several times on assault-and-battery charges. He had some sadistic compulsions, which he took out on animals by starving them and having them fight to the death.

In 1955, he had been arrested for fondling a young girl, but the charge was dropped. That year, his first child was born and she had congenital pelvic disease, which had a significant impact on DeSalvo's home life. His wife, terrified that she would have another child with a physical handicap, avoided sex with DeSalvo, leaving him as a disgruntled married celibate. But DeSalvo reportedly had a voracious sexual appetite, which caused him to balk at his wife's decision. He supposedly found other outlets.

Between 1956 and 1960, he was arrested several times for breaking and entering, and each time he received a suspended sentence. In 1960, his son Michael was born without any physical handicaps.

In spite of his encounters with the law, Albert seemed able to keep a job. After working as a press operator at American Biltrite Rubber, he worked in a shipyard and subsequently as a construction maintenance worker. His boss characterized him as a decent family man and good worker.

Upon his arrest in 1964, the thirty-three-year-old DeSalvo admitted breaking into four hundred apartments. And he had assaulted some three hundred women in a four-state area, he said, although given his tendency to self-aggrandize, these figures were quite possibly exaggerations.

On February 4, 1965, DeSalvo was sent to Bridgewater State Hospital for observation. Shortly after his arrival, a man named George Nassar also became an inmate there. He had been charged with an execution-style murder of a gas station attendant, but he was no ordinary thug. His IQ approached genius level. He was put in the same ward with DeSalvo and became his confidant. Nassar, somehow learning that DeSalvo was the Strangler that everyone was looking for, told his lawyer, F. Lee Bailey, of his suspicions.

In early March 1965, Bailey visited DeSalvo with a tape recorder. Not only did Albert confess to the murders of the eleven official Strangler victims, but he admitted to killing two other women, Mary Brown in Lawrence and another elderly woman, Mary Mullen, who apparently had died of a heart attack before he could consummate his attack on her.

Bailey called Lieutenant Donovan of the Boston police and suggested that he might have a suspect for him, but first he wanted Donovan to provide him with some pertinent questions to ask about the skein of unsolved rape-murders that would help to determine if DeSalvo was being truthful.

Armed with his dictaphone, Bailey went to visit DeSalvo a second time. He said of that interview: "Anyone experienced in interrogation learns to recognize the difference between a man speaking from life and a man telling a story that he either has made up or has gotten from another person. DeSalvo gave me every indication that he was speaking from life. He wasn't trying to recall words; he was recalling scenes he had actually experienced. He could bring back the most inconsequential details . . . the color of a rug, the content of a photograph, the condition of a piece of furniture. . . . Then, as if he were watching a videotape replay, he would describe what had happened, usually as unemotionally as if he were describing a trip to the supermarket."

Commissioner McNamara and Dr. Ames Robey, the psychiatrist at Bridgewater, were called into the consultation. There was much legal wrangling while Bailey tried to protect his new client, DeSalvo, from execution. Yet the stakes were high for Attorney General Brooke, who had announced himself to be a candidate for the United States Senate. Solving the Strangler case would boost his public persona and his campaign markedly.

Finally, on September 29, 1965, the interrogation was complete. More than fifty hours of tape and two thousand pages of transcription resulted. While each detail of the confession was checked out, Bottomly, Brooke, and Bailey (the "B boys") tried to work out what would happen next.

Initial doubts about whether DeSalvo was the Strangler quickly dissipated for the task force. He seemed to know too many details about the crimes and crime scenes to just dismiss him as a publicity monger. Eventually, the Strangler Bureau came to the same conclusion that F. Lee Bailey had—Albert DeSalvo was their chosen one, the Boston Strangler.

That is how he has been portrayed in the public consciousness for four decades, including the highly regarded, at least cinematographically, Hollywood potboiler with actor Tony Curtis as the Boston Strangler. But there were a number of problems with this general acquiescence in the police view that Albert DeSalvo, without the stigma of a conviction for any of these crimes, was the Boston Strangler. These perturbations inspired author Susan Kelly during the 1980s to take up the loose ends of the investigation and to see where they led. Her own book, *The Boston Stranglers: The Public Conviction of Albert DeSalvo and the True Story of Eleven Shocking Murders*, gives the details of her probing investigations and the conclusions she derives from them.

FIRST, THERE WAS THE ISSUE OF DESALVO'S MOTIVE TO CONFESS. Current exonerations through DNA profiling have shown quite dramatically that people do confess to crimes that they have not committed but are convicted anyway. Was Albert DeSalvo's confession of that false character?

Many persons believed that since he was in prison anyway and promised an immunity from the death penalty, he figured he might as well profit while confined. Apparently, he was under the mistaken impression that the reward money of $10,000 was for each victim, which would amount to over $100,000—a tidy sum to leave to his family. Once he confessed, there was also talk of book, magazine, and movie deals that could further enrich him. In today's world of criminal law, such profiteering from book deals by convicted criminals is widely proscribed by statute, but that was not the case in 1964.

Another motive could have been simply his craving for notoriety. DeSalvo was characterized as a braggart of the worst sort. He fancied himself the most notorious figure in the prison. It would not be the first time he took credit for crimes he did not commit. He had once said he was responsible for an assault and robbery in 1964 in Rhode Island, but the victim identified someone else.

Dr. Robey, the prison psychiatrist at Bridgewater, reported that "Albert so badly wanted to be the Strangler." He also said that DeSalvo showed an undifferentiated schizophrenic pattern in his personality. He had a tremendous insecurity and a need to identify himself as a notorious character. He craved attention. He also had an overwhelming compulsion to confess, with no appreciation for the consequences of what he might say. He just needed to be famous, however he might achieve that goal.

If Robey's views were on target, they could just as well explain why DeSalvo might have committed these garish and grotesque murders attributed to the Strangler. Serial arsonists are, for example, well known to set fires in pursuit of the media attention attendant upon and generated by their criminality.

But even if these motives undermine the authenticity of his confession, the question remains as to how he managed to recall so many details of the crime scenes. Robey had an answer for that as well. DeSalvo admitted that he was so fascinated by the press accounts of the murders (with addresses) that as an experienced breaking-and-entering man, he was able to gain entry into the apartments and look around. Apparently, he also had a photographic memory.

Robey cited an example of how he had tested this hypothesis: "We had a staff meeting [at Bridgewater] with about eight people. Albert walked in and walked out. The next day we had him brought back in. Everyone had on different clothes, was sitting in different positions. I said, 'Albert, you remember coming in yesterday? Describe it.' Albert did, perfectly."

In addition, there were many sources available for DeSalvo to learn the minutest details of the crimes. The *Record-American* had printed up a chart, along with the victims' photos, called "The Facts: One Reporter's Strangler Worksheet." It summarized the salient facts of each crime, what the victims were wearing, their hobbies and affiliations, et cetera. Author Susan Kelly has concluded, "That DeSalvo had memorized this chart is apparent because in his confession to John Bottomly, he regurgitated not only the correct data on it but the few pieces of misinformation it contained as well."

There were also rumors that DeSalvo had absorbed information from someone at the prison who was the real killer (possibly his prison mate Nassar), and that DeSalvo's questioners (who had much to gain by closing the case) had led him to the answers they sought.

Yet even with all of these prime sources for factual authenticity, one can find multiple errors in DeSalvo's recollections. For example, he did not recall that he had stabbed Samans seventeen times. He mentioned five or six. That's quite a discrepancy. He also did not know that he had tied anything around her neck (two stockings and a scarf). There were others as well, but my primary interest was how accurate or mistaken he was in his confession to killing Mary Sullivan.

Some officers on the task force were not convinced that DeSalvo was the perpetrator, but they had not been allowed to ask him questions. They had suspects for some of the crimes who were better candidates than DeSalvo, so they thought. Detective John Moran, who was on the Evelyn Corbin case in Salem, took this position. He said later that some of DeSalvo's statements were not credible. For example, DeSalvo claims that he told Ms. Corbin that the "super" had sent him, which is a term not used in Salem. Also, she was familiar with the building's handyman, so she would not have believed this introduction, especially on a Sunday.

DeSalvo claimed to have come in the front door, but Moran believed that the killer had come in from the fire escape, where a doughnut had fallen from his pocket. He thought the real killer was a psychotic man from nearby Lynn, Massachusetts, who had a record of assaulting older women and who had bought doughnuts that morning.

Similarly, author Susan Kelly found other likely suspects for some of the murders. The most significant one was George Nassar. At age seventeen, he had committed his first murder. He was pleasant and well-mannered on the surface, but cunning and unprincipled underneath. It was rumored he was responsible for at least seventeen deaths, and he had admitted that he killed for excitement.

On September 29, 1964, he had murdered a gas station attendant during a robbery, execution style, and attempted to kill a female witness and her daughter. He was sent to Bridgewater State Hospital, where he spent many hours in conversation with DeSalvo. It was Nassar who approached F. Lee Bailey to represent DeSalvo. Yet two women who believed they had encountered the Strangler thought that Nassar looked more like the man they had seen than DeSalvo did.

There was another suspect in the deaths of Slesers, Nichols, Blake, Irga, and Sullivan. He had tried to kill his mother, whom he regularly punched and kicked. He was psychotic with a low IQ. He was identified as the attacker of two young women, and was missing from Boston State Hospital on the days when the five murders took place. He reportedly told his sister Claire that he was the Strangler. His own psychiatrist suspected that he might be the killer. He was eventually transferred to Bridgewater, and his stay overlapped that of DeSalvo.

Nobody who knew DeSalvo believed that he was the Strangler: his wife and family, his former employers, his lawyer, an eminent prison psychiatrist, and even the police who had become familiar with Albert from his arrests for breaking and entering. The lawyer who took over his case after DeSalvo fired Bailey said he did not believe that DeSalvo was the Strangler.

Susan Kelly writes that there was "not one shred of physical evidence that connected him [DeSalvo] to any of the murders." Nor could any eyewitness place him at or near any of the crime scenes. Albert had a relatively memorable face, particularly because of his prominent, beaklike nose. People who believed they had seen the Strangler did not identify DeSalvo as the person whom they had seen.

In addition, three fresh Salem cigarette butts were found in an ashtray near Mary Sullivan's bed. Neither Mary nor her roommates smoked this brand. A Salem cigarette butt was also found in the toilet at 315 Huntington Avenue the day Sophie Clark died there. DeSalvo did not smoke.

Oh, if only DNA profiling had been on the crime laboratory scene in the 1960s. At the very least, the saliva, with smoker's DNA molecules, could connect or not the cigarette butts from one crime scene to the others. Also, the DNA from the cigarette butts could be compared with that of DeSalvo and the other likely suspects, eliminating Mary and her roommates as the person who had smoked the discarded cigarette.

Yet, even in the sixties there were scientific tools that could have been called to the task when the cigarette butts were found at two different Strangler crime scenes. Saliva not only possesses the DNA marker of the salivator, it also contains the A–B–O blood antigens of those 85 percent of the population who are secretors (persons whose A-B-O blood antigens are found in their nonblood bodily fluids, like saliva).

Thus, if DeSalvo was a nonsecretor, then finding a secretor's A-B-O blood type on the cigarette butts would be dynamite evidence of De-Salvo's nonpresence and noninvolvement in the crime, assuming Mary's murderer was the smoker who left the butts behind.

Further, even if there was not found to be a disparity between the secretor status of the saliva on the cigarette butts and that of DeSalvo, there could be a difference in the A-B-O markers (for example, DeSalvo could be type B and the saliva type A), pointing to DeSalvo's not being associated with the cigarette butts.

Finally on the topic of disparities, the victims and the MO for the various crimes were grossly dissimilar. Kelly summarizes some of the more obvious differences:

No similarity whatsoever exists between the relatively discreet killing of Patricia Bissette, whose murderer tucked her into bed, and the ghastly homicidal violation inflicted on Mary Sullivan, in which she was degraded by having a broom handle shoved into her vagina either after or, horrors to behold, before her death. Beverly Samans was stabbed but not sexually assaulted. Joann Graff was raped vaginally and strangled. Evelyn Corbin performed—probably under duress—oral sex on her killer. Jane Sullivan (no connection to Mary Sullivan) was dumped facedown into a bathtub. Ida Irga was left in the living room with her legs spread out and propped on a chair. Whoever the Boston Strangler was, his methods of exhibiting the bodies of his victims were very individualistic and lacking in comparability except for being a demonstration of his exhibitionist's barbarity.

DESALVO NEVER STOOD TRIAL FOR ANY OF THE MURDERS. HIS confession was inadmissible in court because there had been no independent corroboration, and there was no physical evidence against him, which was admitted in a confidential memo from the Attorney General's Office. So Bailey worked out a deal with the Commonwealth to ensure that his client would not face the death penalty. He defended DeSalvo in his trial for the Green Man crimes, hoping to prove him insane. But the jury convicted him on all counts, effectively denying him a lifetime stay at a forensic hospital. He was to go to Walpole State Prison.

An inmate with whom DeSalvo later engaged in a foiled escape attempt said that DeSalvo denied being the Strangler and for his "confession" had been tutored in the crime scenes. This man claimed that he'd overheard fifteen to twenty such conversations with another inmate at Bridgewater, but did not name the other man.

In November 1973, as Albert DeSalvo was serving out his life sentence at Walpole State Prison, he was stabbed to death in the infirmary. Kelly writes that on the evening before he was murdered, he telephoned Dr. Ames Robey and asked to meet with him. DeSalvo was frightened, Robey later reported, and said he had something important to reveal. Robey arranged to meet with him the next morning, but that meeting never occurred.

Based on a meeting he had tried to set up at an earlier time with Robey and a reporter, Robey believed that DeSalvo was going to confess about not being the Strangler. As we learn from Kelly's book, Robey explained, "He was going to tell us who the Boston Strangler really was, and what the whole thing was about. . . . Something was going on within the prison, and I think he felt he had to talk quickly."

Although no one could corroborate Robey's story, DeSalvo's lawyer, Thomas Troy (F. Lee Bailey's replacement) said that DeSalvo had recanted in his presence, hinting that he knew who the Strangler really was.

The manuscript that DeSalvo claimed to have been writing about himself when he was killed was never located, nor were his killers ever successfully prosecuted, although the only surviving suspect among them is now lodged in the Maryland penitentiary system.

IN 1999, THE BOSTON POLICE DEPARTMENT ANNOUNCED THAT IT was seeking evidence from DNA analysis to try to resolve the controversy over whether or not DeSalvo was indeed the Boston Strangler. The Cold Case Squad had looked through the boxes of evidence from more than three decades earlier and decided that if they could come to an unshakable conclusion about this famous (or infamous) case, they could shine the piercing light of a laser on the real possibility of solving old cases. Yet some of the evidence was beyond their reach. The knife used to stab DeSalvo, which would have contained his DNA, was missing, as were biological samples taken from some of the victims.

Also in 1999, Michael DeSalvo, Albert's son, requested DNA testing as a way to resolve this case and perhaps exonerate his father as the notorious killer. Enough had been written about DeSalvo's motives to confess falsely, and about the lapses of memory evident in the contents of the confession, that there was now doubt aplenty about the truth of his confession.

However, the Boston Police Department issued a statement that the foreign biological fluid samples they had from the victims were not adequate for conclusive testing.

At a dead end with the authorities, Diane Sullivan Dodd, through her son, Casey Sherman, asked me to investigate Mary Sullivan's death. Ms. Dodd and Mr. Sherman strongly believed that the identity of the person who had killed Mary Sullivan was still a legally unresolved issue. Casey Sherman's recent book, *A Rose for Mary*, thoroughly canvasses the controversy over DeSalvo's being the murderer.

Concerted and extensive efforts were then conducted to obtain any and all reports on the various official state and local investigations that had presumably been launched into the death of Ms. Sullivan, both through researching newspaper and other archival materials as well as via written communications to a number of officials and official agencies in Massachusetts. Aside from the public reports of the Boston Strangler Bureau, organized by then Attorney General Edward Brooke, all such efforts to obtain firsthand information were blocked by the authorities.

I faced the continuing claim by Massachusetts officials that the death of Ms. Sullivan was an open, albeit cold, case still officially and theoretically under investigation. In other words, they did not have to turn anything over to us since their investigation was ongoing. Shades of Huey P. Long's resurrected fifty-six-year-old investigation by Louisiana police when and only when I launched my own investigations.

Notwithstanding, I did obtain one item of vital importance, courtesy of an investigative reporter: a copy of the official report of Dr. Michael

Luongo, resulting from his January 5, 1964, autopsy of Mary Sullivan. We found it both enlightening and disturbing. When compared to DeSalvo's confession to the killing of Ms. Sullivan, the report revealed major discrepancies between what DeSalvo said he had done and what the autopsy said had actually happened to Mary Sullivan.

The first disparity related to her estimated time of death. The report indicated that Dr. Luongo had arrived at 44A Charles Street to view the deceased's body at 7:30 P.M. on the day of Mary's death. At that time he noted the presence on the body of "full rigor mortis, and marked dependent lividity," whereas Mr. DeSalvo had confessed to having entered the premises (there being no signs of forced entry) at 4:00 P.M. the same day.

It goes without argument that full rigor would be most unlikely some three and a half hours later, unless there were, in pathological terms, very unlikely and uncommon circumstances causing a rarely encountered instant rigor or cadaveric spasm. In the trenches in France, for example, in the First World War there were reports of soldiers found dead with a coffee cup tightly clasped in their hand, a rare phenomenon that could be attributed to instant rigor.

But normally, after three hours some small muscles may have stiffened, but full rigor takes eight to twelve hours or even longer to become fixed. Moreover, the stomach contents at autopsy were given as "contains 2 or 3 ounces of brownish fluid and mucus. This has a slight odor of coffee." The small and large bowels were essentially empty. If she was killed at 4 o'clock, that was a remarkably long time for Mary to go without food. These findings, taken together, indicate a time of death in the morning, after a cup of coffee and many hours before DeSalvo claimed to have entered the apartment.

DeSalvo stated in his confession that he had "intercourse with her," repeating that phrase twice. Even though he did not admit to having ejaculated during the intercourse, it would have been expected due to the other sexual perversions inflicted on Ms. Sullivan, during which he did claim to ejaculate. The autopsy report stated that "no spermatozoa are found in smears of the vaginal and rectal areas," although semen was detected draining from her mouth to her exposed breasts.

As in the case of the first victim, Anna Slesers, where DeSalvo claimed he had inflicted blunt-force trauma, and Dr. Curtis found such evidence at that autopsy, DeSalvo described hitting Mary Sullivan with his fists, in which "I almost knock her out" after battering her "on the face and on the breasts and on the belly." In contrast, Dr. Luongo reported "no edema or external evidence of injury" to Mary's skull. However, he observed "acute traumatic injuries of both breasts"—injuries that could have been bite marks inflicted by DeSalvo, as he claimed to have done in the killing of Ms. Slesers.

It is important to note for the factual record that Albert DeSalvo was, during his tour of duty in the U.S. Army in Germany, a prizefighter with a barrel chest and broad, muscular shoulders and an athlete's torso to match. In short, if he hit Mary Sullivan with the battering-ram force that he said he did, the evidence of it should have been as plain as a pikestaff at her autopsy, but it wasn't.

From DeSalvo's confession, it could be understood that he claimed to have caused her death by manual strangulation. Yet the autopsy report indicated that her death was the result of asphyxia due to strangulation by ligature. The hyoid bone in her neck had not fractured, as would have been likely given the type of manual strangulation that DeSalvo had described. A nineteen-year-old's hyoid is more resilient and malleable than that of a man of my seventy-odd years; still, at any age, with the right amount of force the hyoid can be fractured or at least traumatized.

Apart from these contradictions and inconsistencies between the autopsy report and DeSalvo's confession, the account he offered included another feature that is at once both incredible and inconceivable. In it, he claimed that during the assault upon Ms. Sullivan "she had in her hand all the time" a little knife, "like you use for peeling vegetables in the kitchen," and that "she did not once lift that knife against me."

That statement, the discrepancies between the autopsy report and DeSalvo's confession, our inability to obtain official information, and our being frustrated at every turn by the Massachusetts law-enforcement agencies in seeking access to possibly still-extant and inventoried evidence convinced Diane Dodd and Casey Sherman that the only recourse for obtaining fresh evidence on the perpetrator of the murder of Mary A. Sullivan was to authorize me to exhume her remains.

I wholeheartedly agreed with the Sullivan family's assessment of the situation. Nevertheless, I recognized that finding such evidence in or on the remains of Ms. Sullivan, some thirty-six years after her death, autopsy, and embalming, was going to be a shot in the dark, with failure more probable than success.

But regardless of the risks of failure, how could I reject the sincerity of Mary Sullivan's long-suffering sister and her sister's son? Maybe the Massachusetts authorities could heartlessly and expediently turn a deaf ear to their searching entreaties, but I lacked the steely insensitivity to do the same.

Yet in view of DeSalvo's claimed sexual assault of Mary Sullivan, which was confirmed in part in Dr. Luongo's report, the remote possibility existed that a DNA-extractible seminal stain might be discovered on tissues obtained from the reautopsy. The logistical aspects now had to fall correctly into place. In April 2000, I received a written consent from Diane Dodd authorizing the exhumation and reautopsy by me and members of my scientific team, with Dr. Michael Baden to be the lead pathologist. Meanwhile, Richard DeSalvo graciously volunteered to provide the necessary reference samples for mitochondrial DNA analysis, utilizing buccal swab and fingerstick kits provided by Tri-Tech, Inc. He was convinced, as much as Casey Sherman was on his part, that his brother was not the Boston Strangler and hoped that my current investigation would prove him right once and for all, at least as to the death of Mary Sullivan. As he put it, "If Albert did not kill Mary Sullivan, then who is to say he killed any of the other women?"

Once again, as in the identification of the remains of Jesse James, I was cheek by jowl with mitochondrial DNA (mtDNA) testing as opposed to nuclear DNA, the kind that is contributed by both parents equally to each of their offspring. MtDNA, it is to be recalled, passes entirely and only from a mother to her children, both male and female. And the female children, and only the female children, will pass their mother's mtDNA to their progeny, and so on through the generations in an exclusively matrilineal clubbishness.

For those who have an insatiable craving for the minutiae of molecular biology to account for the mother's contributing all of her mtDNA and the father none of his to their children, the resolution to this biological riddle is readily explicable. A spermatozoa has two parts: the head, where the father's nuclear DNA is lodged, and tail, which houses the mtDNA. When a sperm impregnates the ovum, it is the head of the sperm that does the business of fertilization, not the tail. The tail keeps the sperm moving, giving it motility, but when the target is reached the tail drops off and the father's mtDNA goes with it.

The mother's ovum thus does double duty, genetically speaking, by providing all of her mtDNA and half her nuclear DNA to the fertilized egg, and the father, in his turn, shares his half of his nuclear DNA from the head of the fertilizing sperm—an equal pairing of the mother's and the father's nuclear DNA, showing nature at its egalitarian best but quite the contrary as to the mtDNA. But the father's mtDNA from his sperm's tail is lost in this impregnation shuffle.

I have described the mtDNA generational transfer process as "nature's preference for the distaff side." That remark generally gets a rise from those of my students sufficiently literate to know that the distaff side refers to women.

Most assuredly, I am right in coining this phrase, for the mitochondria are the organelles that function to produce the energy for our body. In a word, the mitochondria are the fuel that drives the body's functions. So if you are an energyless couch potato, blame your sedentary condition on your mother's mtDNA. Just a sad biological joke, but it makes a hit among my students, at least the A students among them. And that's the boost in energy for which all professors strive.

The last legal step before I would have touched all the bases for the exhumation was to be certain that all statutory requirements had been fulfilled. Strangely, even though Massachusetts was the first state to adopt a statutory framework for a medical examiner's system, well in advance of Jack Klugman's popularizing the subject in the *Quincy* television series, there were no statutory formulae governing the necessary legal authorization for the exhumation I envisioned conducting.

However, a board of health permit was said by the superintendent of the St. Francis Xavier Cemetery, where Mary was interred, to be all that he required, so long as it was issued at the request of a family member. I made arrangements to journey to the city hall in Barnstable in company with Diane Dodd and Gaetan Cotton, my photographer, to secure the permit.

It did not hurt matters that Diane had a high school classmate working in an administrative capacity at the city hall. The only hitch arose with a stipulation on the exhumation form to list not only the grave from which Mary was to be exhumed but also the grave in which she was to be reburied. I knew they were to be one and same, but the form did not make provision for noting that eventuality.

A curious clerk wanted to know why the removal and the reburial were to be from and to the same Sullivan family plot. Without mentioning the name Albert DeSalvo, I put on my most serious professorial mien and pointed to the existence of a deeply rooted tradition in cemetery burial practices that a child should be buried on the heart side of his or her deceased mother. It was simply to fulfill that tradition that this exhumation was organized. I drew to her attention that Mary had predeceased her mother, giving rise to the present need to straighten things out to comply with the tradition.

The desk clerk was not completely convinced (taken in?) by my rationale, as evidenced by the frown that crossed her face. Changing her tack, she looked at my baseball cap, which publicized my affiliation with the GWU Forensic Sciences Department, and quite bluntly and boldly asked, "Who are you, anyway?"

Without dropping a beat, I replied, "One of the Sullivan clan."

That answer having saved me from further questions, she gave the signed permit to Diane and we left the building. Outside in the treed yard fronting the building, Diane looked at me with a twinkle in her Irish eyes, as much as to say it was only a small lie, after all, and for a good cause.

Seeing her amused skepticism, I put her at ease by pointing to my own Irish heritage. "My mother," I told her, "was in truth a Sullivan. Therefore, I am one of the Sullivan clan, am I not?"

She just gave me a full smile, and we walked away as more than friends.

THE OBJECTIVES OF THE PROJECT WERE TWOFOLD: (1) TO SUBJECT the physical evidence secured by law enforcement in 1964 from the person of Mary Sullivan (if we could obtain it) to scientific technologies not available in 1964, and (2) to examine her remains for evidence that might shed light on whether Albert DeSalvo had raped and murdered her.

As with my other exhumations, I organized a scientific team of experienced and acclaimed experts, along with a support staff of qualified specialists, which included my daughter-in-law, Traci Starrs. Among them were Michael Baden, M.D., the former chief medical examiner of New York City; Henry Lee, Ph.D., the former chief of the Connecticut Forensic Science Laboratory; Major Tim Palmbach, current head of the Connecticut State Police Forensic Science Lab; David Foran, Ph.D., the director of the Forensic Molecular Biology Laboratory of The George Washington University; and Todd Fenton, Ph.D., a forensic anthropologist with Michigan State University, who had been one of my team in the Packer exhumation.

They would provide services for the reautopsy, crime-scene analysis, and DNA analysis. We also had a radiographer, a videographer, and an odontologist. Other disciplines, such as toxicology, were represented by experts at a distance and awaiting our submission of evidence to them, such as Bruce Goldberger, Ph.D., a toxicologist with the University of Florida Medical School who had been a toxicological mainstay in my Frank Olson/CIA investigations.

From what I learned about the circumstances of the 1964 burial, Mary Sullivan had been embalmed and then placed in a wooden coffin, concrete vault, and bricked-in liner. I asked George Stephens, Ph.D., of The George Washington University, who was my in-house forensic geophysicist, to perform the requisite geological analysis of soil samples from the grave site. His analyses showed the soil to be sandy and well-drained and permeable, all of which was conducive to well-preserved remains.

The opening of Mary Sullivan's grave and the exhumation commenced on Friday, October 13, 2000, at the St. Francis Xavier Cemetery in Hyannis, Massachusetts. The weather was clear and turned out to be quite comfortable.

A camera crew from the CBS television program *48 Hours* was also present. But we steadfastly kept our exhumation plans secret from any other members of the media in deference to the wishes of the Sullivan family.

A backhoe removed the well-drained, sandy topsoil from the monument-marked grave of Mary Sullivan, and we saw the tripartite lid, with eye hooks attached, of a concrete grave liner. The backhoe was back in action to lift off the section of the concrete lid covering the head end, disclosing a collapsed wooden coffin underneath.

Water could be seen at the bottom of the coffin. The skull, facing skyward, was plainly visible. Yet the face seemed to be remarkably preserved for a thirty-six-year-old burial, giving us all a surge of optimism as to what we might find.

In view of the fact that a body bag would be needed to remove the remains from the fragmented coffin, we decided to replace the lid and return the next day to complete the exhumation.

On October 14, upon the removal of the three parts of the concrete vault lid, it was immediately evident that Mary's remains were almost fully skeletonized, with only fragments of tissue attached like a thin layer of firm plasterlike cardboard. The casket was lifted out gingerly, revealing that the face, which the previous day had seemed well-preserved, was actually covered with artifactual, castlike material creating a masklike impression—probably a consequence of the mortuary's preparation of the remains for a viewing by the family. The remains still in the casket were conveyed to the John Lawrence Funeral Home in Marston Mills, Massachusetts, where a mortuary suite had been graciously provided free of charge for the reautopsy. Mitchell Calhoun diligently recorded the entire procedure on videotape, documenting every jot and tittle of our actions.

As we laid Mary Sullivan's remains out on the table, I saw that her hands held rosary beads. Before going further, I scanned her body with a portable ultraviolet light source, looking for a luminescence indicative of the likely presence of seminal fluid, but not specifically semen. The UV scanning was followed by an alternate light screening by Tim Palmbach, who came from his laboratory well-provisioned to assist us. His runs corroborated what I had already discovered with the UV light, showing possible traces of semen both in Mary's head and pubic hair as well as in the underpants in which she had been buried.

We had a complete skeleton, consistent with thirty-six years of interment. Grave goods, as well as the match of mtDNA control-region sequences to the deceased's sister from a bone fragment, confirmed the body as that of Mary Sullivan. The abdominal organs and brain were situated inside a plastic bag that had been placed in the abdomen after the autopsy, but unlike other nonskeletal parts of the body, these showed a remarkable degree of preservation. There was no evidence of breakage in the bag. They were readily identifiable as distinct organ structures, and samples were taken for toxicological studies by Bruce Goldberger at his University of Florida Medical School's laboratory. In my many exhumations, Bruce had been a mainstay, with thorough tox screens, comprehensive reports, and audiovisual slides to demonstrate his conclusions.

However, David Foran's later DNA testing at his GWU lab could not isolate typeable DNA from these tissues. The explanation, in light of the pristine condition of those tissues, eluded us. Since Bruce Goldberger found no preservative in the tissue samples submitted to him, the lack of DNA for profiling could not be placed at the foot of a preservative's degradation of the DNA. Fortunately, the DNA in the tissues was not central to our quest—just another unanswerable scientific conundrum to ponder over the long haul.

With DeSalvo's confession to killing Mary Sullivan keenly in mind, a number of our autopsy findings were a challenge to the accuracy of the confession. First, Todd Fenton reported that there was no sign of trauma to her skull, despite DeSalvo's saying he'd knocked her out with a blow to the head. Yet our particular interest was directed to the neck organs, since DeSalvo had claimed to have manually strangled Mary with three ligatures, as confirmed by Dr. Luongo's autopsy report, as well as with his bare hands. The hyoid was found to be unfractured and otherwise intact, as well as completely fused, a somewhat early fusion for a nineteen-year-old.

All samples retained for subsequent examination were given access numbers and entered sequentially in a logbook by Traci Starrs and Michelle Hamburger, two stalwarts in the field of documentation of specimens. All sample containers were marked with the same logbook entry numbers, including brief descriptions of the contents. All in all, some sixty-four individual items were tabulated, many of which would require future laboratory testing.

Our ultimate goal was to test any and all evidence as thoroughly as possible to see if Albert DeSalvo might in any way be identified, or even suggested, as her murderer. Of course, the possibility of there being evidence exonerating him was also front and center in our vision. The results that would tighten the noose of guilt around Albert DeSalvo would have been to find DNA consistent with him on Mary Sullivan. The controversy over his involvement could perhaps then be halted.

The next morning, Sunday the fifteenth, we returned to the funeral home to prepare Mary Sullivan's remains for reburial. I was with Diane Dodd and a reporter from the Boston *Globe* who had learned of the exhumation. I had noticed that the time for the last mass at St. Francis Xavier Church in Hyannis was 12:00 noon. I was in hopes of finishing in time to attend that mass, but as it turned out, more compelling matters intervened.

George Stephens and I were alone with Mary's remains in the funeral home, preparing her for burial, with the family and others waiting outside. Time was running short, but I could still make the noon mass.

After having placed a white shroud over Mary's remains in the new coffin for her reburial and before placing the lid on top, I hesitated. "George," I asked, "do you think Mary's sister might like to say a last good-bye to Mary?"

"Absolutely," he replied. George has always been a man of few, but choice, words.

We walked together outside, where I offered Diane the opportunity to visit with her older sister before we closed the coffin forever. She accepted gratefully, and while she was alone inside with her sister I waited as the clock ticked closer toward noon.

Shortly after noon, Diane exited the funeral home, making no effort to conceal the tears staining her cheeks. Before reentering the mortuary with George to nail the coffin lid permanently in place, I made a most regrettable blunder.

I looked at Diane and in a voice of some dismay said, "Well, I guess I've missed mass."

With quiet compassion Diane said, "I think God and Mary will forgive you."

Never have I felt so completely chastened.

The reburial was next on our agenda, with my team members in attendance at the St. Francis Xavier Cemetery, some of us taking a role as pallbearers. In a quiet and dignified service, a Catholic priest offered prayers for Mary's immortal soul and asked God to give peace and comfort to

Mary's grieving friends and relatives who had suffered through two burials of her remains.

The exhumation immediately became worldwide news. Now the pressure was truly on the Massachusetts authorities to do the right thing by the family. It soon came out that they had suddenly and unaccountably found six semen samples from the original Sullivan investigation and would be testing them and, in the course of doing so, necessarily destroying them.

I waited for a call from the Boston police lab. The lab was to share with me an aliquot of the six semen samples they had discovered, and I was to share with them reference samples from Mary, Diane, and the DeSalvos to compare against whatever DNA test results they might obtain. Not unexpectedly, the call never came and hasn't to this day.

Seemingly, the official hands-off policy toward my investigation was still in place and had become implacable. Once again I had been foiled by my own incorrigible scientific naïveté. I had been schooled in the notion that science was rendered as an open book, with its processes and conclusions subject to public and peer scrutiny. Apparently, the Boston Police Crime Laboratory was not of a similar scientific bent.

THE MATERIALS SUBMITTED FOR DNA ANALYSIS TO THE MOLECU-lar biology laboratory of Dr. David Foran at The George Washington University included a segment of head hair and another of pubic hair, eyelashes, fingernails, viscera, and underpants, as well as soil, small coffin fragments, and other miscellaneous items from the exhumation.

In my frequent visits to Dr. Foran's lab, playing the worried Jewish mother as I did, I noted that extreme care was being taken when handling the evidence at the lab in order to minimize the chance of contamination. Thus, the two sections of the underpants that gave presumptive evidence of semen were processed separately to avoid cross-contamination. As was our procedure during the autopsy at the John Lawrence Funeral Home, all who were in contact with or in the same place as the samples to be tested wore gloves, gowns, and masks, and before use all solutions and supplies were sterilized and subjected to short-wavelength ultraviolet irradiation.

Preliminary examination of the evidence included a complete inspection using a long-wavelength ultraviolet light source. Areas that showed fluorescence underwent further analyses.

The target areas of fluorescence, the head and pubic hair and the underpants, were small in number as well as small in size. In recognition that mtDNA testing is destructive of the samples under examination, Dr. Foran eschewed tests that would have been nonspecific identifiers of the samples while at the same time destructive. But the microscope, being nondestructive of the samples, was not of concern in this regard.

The optical light microscope was used to examine the items first to see if a most unlikely event had occurred—namely, that spermatozoa had survived, either as heads or separated tails, all these thirty-six years since Mary's death. They were, as anticipated, nowhere to be seen.

Second, the magnification under the microscope could make it plausible, from the translucent clumping of the material, that the items had the appearance of dried, crusted semen stains. Indeed, such was found to be demonstrated by the material intertwined in the pubic hairs.

All the laboratory work included controls and blanks, with replication of the test results mandatory to guarantee their reliability. Moreover, a very conservative approach was adopted: If bands appeared in the blanks or controls, the entire experiment was voided and a new try commenced.

None of the reference samples from Richard DeSalvo were tested or even maintained in the same facility where the unknown samples from the grave were being processed. The lab staff was scrupulous in their concern to avoid contamination and in their efforts to avoid any bias in the conduct of their tests.

If we had sequenced Richard DeSalvo's DNA first, who was to say that upon later sequencing the DNA from the specimens from Mary's remains we would not be, even unconsciously, on the qui vive for what we already knew to be Albert's qua Richard's mtDNA? Why take such a gamble with the integrity of our work product? Do the unknown samples come before the known samples in the proper order of scientific business, not the reverse?

The mtDNA process is by no means an offshoot of today's fast, even quicker-than-fast, culture. It was Nobel Prize winner Kary Mullis who paved the way for the mtDNA testing that has become a staple of law enforcement and clinical laboratories. Mullis recognized the basic fact that enzymes can unzip the two sides of the helical structure of DNA. His technique advanced this knowledge by using enzymes to multiply some amounts of DNA to the point that there would be a sufficient supply to be analyzable. He took the DNA of a sample and enlarged it so that it could be meaningfully examined. But his system could not shortcut around all of the laborious aspects of mtDNA profiling.

As a consequence, untold hours of uncompensated but satisfying labor by Dr. Foran and his students were poured into the extraction of mtDNA, the amplification through Mullis's PCR (polymerase chain reaction) procedure (sometimes called chemical xeroxing), the sequencing of the mtDNA in the search for the control regions, the interpretation of the results, and the population statistics to be factored into the results. Aside from the amplification, which is automated, so much of the rest of the processing is exceedingly labor-intensive and, therefore, challenging and time-consuming.

Dr. Foran and his team's results were published in a peer-reviewed article, jointly authored by him and by this book's primary author, in the January 2004 issue of *Medicine, Science and the Law,* "the official journal of the British Academy of Forensic Sciences." I will not detail the specifics of Dr. Foran's laboratory procedures (which appear in full in the article), refusing to emulate many of Dorothy Sayers's Lord Peter Wimsey mystery novels, where the mystery of the chemistry of poison analysis often bulks larger than solving the mystery in question itself. Suffice it to say that Dr. Foran's team's results were unquestionably phenomenal, opening not a new page but a new book on the Boston Strangler murders.

The mtDNA testing first established that the remains were in fact those of Mary Sullivan, even though Todd Fenton had already done so anthropologically using pictures of Mary's smiling face in life and a superimposition of them on the skull from the coffin. In addition, Mary's smile showed that one of the incisors in her upper teeth bore a unique configuration that the coffin's skull replicated. Thus, the twin disciplines of anthropology and odontology declared the remains to be those of Mary, but I was not satisfied without the conclusive confirmation of mtDNA.

At the autopsy in Massachusetts I obtained a bone section from Mary, which was to be pulverized and the mtDNA sequenced to be compared to the reference sample I had obtained from Diane Dodd, Mary's sister. Dr. Foran did the mtDNA matching, which proved we were truly analyzing the remains of Mary Sullivan.

The identification of Mary through her mtDNA was imperative to ensure that none of the test samples from her body were in fact the product of her mtDNA rather than her attacker's or another person's. Our lab's mtDNA testing also had to exclude all of my team members who were in contact with any of the evidence, which most particularly included Dr. Baden, Major Palmbach, Dr. Fenton, Traci Starrs, and me. We were, happily, excluded.

I had planned to present our findings to the annual meeting of the American Academy of Forensic Sciences in Seattle in February 2001, but I was upstaged by Chief Federal District Court judge William Young. Judge Young was presiding over litigation in his federal court in Boston, brought by the DeSalvo family and Diane Dodd and Casey Sherman to secure access to any evidence relevant to Mary Sullivan's death still in the custody and control of Massachusetts law-enforcement authorities.

As I stood at the podium before a packed house of forensic scientists at our Seattle meeting, about to announce our results, with my team members ready to join me with their individual findings, I was stunned to learn from Elaine Whitfield Sharp, the DeSalvo family's lawyer, that Judge Young had issued a gag order against any public disclosure of our work product. It was a rerun of Louisiana judge Ortique's gag order in my investigation into the death of Huey P. Long.

This time, Judge Young being a federal judge with far-reaching jurisdiction, I decided against going forward and chancing a contempt-of-court citation. But I did tell the assembled audience that our findings were dramatic and even galactic in their significance. I left the specifics to a later date, which, however, had to await my exhuming the remains of Albert DeSalvo himself.

In the meantime, I had received the DeSalvo family's consent to exhume Albert DeSalvo's remains and perform an autopsy, so my team went back to Massachusetts for another exhumation. It was hoped that evidence from his body would answer questions about both his murder and that of Mary Sullivan, at least assuring us that the mtDNA we had tested from his brother Richard was, indeed, his mtDNA. Better to leave no stone unturned—or, in this instance, no grave unopened.

Early on the morning of October 26, 2001, a year after the exhumation of Mary Sullivan, my team gathered in Puritan Lawn Memorial Park in Peabody, Massachusetts, for the second exhumation. With us were author Susan Kelly, who had written *The Boston Stranglers,* a book that put DeSalvo's confession into serious question, as well as DeSalvo family members.

My daughter-in-law, Traci Starrs, a regular and a stalwart in my exhumations, recorded the data as we raised the coffin from where it had been buried for the past twenty-eight years—since DeSalvo had been murdered in prison. Mitchell Calhoun recorded it on videotape and Gaeton Cotton photographed each step of the process for our records.

Funeral directors Kevin Watts and Leo Barry, from the John Lawrence Funeral Home in Marstons Mills where we had done the reautopsy of Mary Sullivan, were also present. In less than an hour, the coffin was raised from the ground and loaded into Mitchell Calhoun's oversized black Ford pickup.

This time, we were taking the remains to York, Pennsylvania, where Dr. Jack Levisky, a forensic anthropologist, had his lab prepared and waiting for us. It was a convenient central location for the rest of the team, which included pathologists Jack Frost and Michael Baden. It was also beyond the reach of the Massachusetts Attorney General's Office. The power of law enforcement being what it was in this investigation, I realized that we could be stopped in our tracks on the same specious claim made in my Huey Long investigation—that we were usurping an ongoing state-sponsored criminal investigation.

A convoy was agreed upon, with a hearse, driven by Kevin Watts accompanied by Leo Barry, in the lead. I was ensconced as a front-seat passenger in Mitchell Calhoun's oversized Ford pickup truck, with Susan Kelly in the rear seat. Gaeton Cotton, our photographic everyman, brought up the rear.

As always thinking ahead, Mitchell had provided walkie-talkies for each car so that we would not stray far from each other while traveling. But Mitchell's pickup needed a gas refill, requiring us to find a gasoline station after crossing into New York.

Like some spastic reptile, we pulled off the main highway onto a two-lane winding road in the search of a gasoline station. We had just passed a military airplane parked off the road to our right in the field of some military installation when, looking up the long hill before us, I noticed that the traffic in our lane was being stopped for a security inspection at the rise of the hill.

My first thought was of Albert DeSalvo in the coffin in the Watts hearse ahead of us, with its coffin fluids still draining out the rear of the car. How could we explain the coffin? How could we explain Albert De-Salvo's presence in it? Would we be required to open the coffin, since coffins have been known as substituting for human "mules" in the transporting of concealed illicit drugs? Did I dare have the convoy do a "Yu-ey," as the Australians describe a U-turn, or would that be certain to bring the security forces after us, the U-turn by the entire convoy signaling a most suspicious move? After all, isn't flight in the presence of law enforcement an indication of guilt?

All my questions to myself were answered by Mitchell, who was on his walkie-talkie directing a U-turn by all of us. Mitchell has always been a man of good judgment and instinctive action. Everybody followed his lead, but I, for one, had my eyes fixed on the rear when, to my dismay, I saw a vehicle speeding down the road toward us. Worse yet, the vehicle had a light on top and it was flashing.

I waited in much discomfort for the inevitable police clap on the shoulder indicating my being taken into custody. Thankfully, it did not happen that way, for the vehicle—a tow truck—was not chasing us but heading pell-mell to the scene of an accident. I don't know why I felt so guilty, for a guilty conscience, according to the well-worn canard, needs no accuser. The incident left us all rollicking with laughter, all the more so for my undeserved unease.

The remainder of the trip was uneventful until we reached York College, where Jack Levisky was waiting for us in the darkness at the rear door to his laboratory building. Kevin and Leo rolled the casket out of the hearse onto a metal gurney, the smell of decomposition hitting those who were close to it, then rolled it into the building, where it was upended so that it would fit into the elevator to the second floor. I suddenly became keenly aware of the logistical difference between an entry to a university's laboratory and the mortuary suite in a funeral home. Universities are not geared to the space requirements of a coffin moving in, around, and about the laboratory. Doorways and elevators are just not suited to the dimensions of a coffin, even in this day of the obese American.

As we moved gingerly from the elevator, the coffin dripped fluid with a pungency that made some in our party pull shirts or sweaters over their noses and mouths. Then we took it down the hall to the doorway of the lab, where it would rest overnight.

But a problem confronted us. The doorway was not spacious and was awkwardly constructed, so we could not upend the casket as we had done in the elevator. We stopped for a moment to think this dilemma through. We could not just leave it there in the hall overnight. With a few Rube Goldberg–like maneuvers we jimmied and cajoled the coffin, still unopened, into the lab. The only hitch continued to be the odoriferous puddles that trailed us on the floor.

In a mood to relax, we all adjourned to dinner. Jack Frost and Major Tim Palmbach had arrived, as had Michael Warren, Ph.D., a forensic anthropologist with the University of Florida's human identification lab. We had a hearty dinner and constructive conversation about the next day's activities, then retired for the night with a general sense of bonhomie.

Joining the team would be Michael Baden, who had been the pathologist on the Mary Sullivan reautopsy; forensic pathologist Patricia Aronica-Pollak to work with Jack Frost; odontologist John McDowell, who waited in his University of Colorado lab for our call; George Stephens, Ph.D., our forensic geologist with a cascade of hash marks from participating with me in these projects; and radiographers Michael and Susan Calhoun. Toxicologist Bruce Goldberger, of the University of Florida Medical School, was awaiting our sending him tissue samples. Attorney Linda Kenney (wife of Michael Baden) and forensic scientists Barbara Hanbury and research assistant Michele Hamburger were also on board. Back at The George Washington University, Professor David Foran and his team of assistants stood by to conduct any DNA tests as necessary. The multidisciplinary scientific team of volunteers was ready, willing, and most assuredly able.

At 8:00 the next morning, the autopsy began with steps preparatory to opening the coffin. The usual equipment had to be arrayed, including evidence bags, photographic equipment, X-ray machines, and autopsy tools.

Finally, it was time to open the lid of the twelve-gauge steel casket. We had fans blowing and windows open, but the ventilation did not entirely eliminate the odor. I like to deceive myself into thinking that I can tough it out and that over time the reaction to the disagreeable smell will dissipate. Not true! I always advise first-timers in exhumation ventures involving fleshed remains (stinkies) to cut their nose hairs and to stuff Vaseline into their nostrils. It is the retention of the smell in the nose that continues after the work is done which is most disconcerting. As for me, I always carry smelling salts in the pocket of my lab coat. It is always good to be prepared.

As we gazed upon the remains of Albert DeSalvo, it was clear that he had decomposed considerably over the nearly three decades since he had been autopsied and embalmed. Albert's trademark nose made it clear that this was our man and no other, but otherwise his face was nearly unrecognizable. Covering the skin was a layer of crusty maggot and beetle husks. The insects had gotten to him. Yet the suit he wore was intact.

The clothing was removed, and in one of the suit coat pockets we found a picture of Albert DeSalvo with an unidentified young man standing beside him. It was barely discernible after all these years, but when X-rayed by Michael Calhoun, Albert's glowingly happy face stood out boldly.

It was time for the pathologists to go to work on the body. They removed several tissue samples, which Traci carefully bagged, taped, and labeled for later toxicological testing by Dr. Goldberger.

To our surprise, the heart, lungs, and kidneys were missing. Generally after an autopsy, these organs are deposited into a bag with all of the other organs and placed into the body's stomach cavity, as had been the case with Mary Sullivan. Who had left those out, and why? That was a mystery we could not settle so many years after the fact. But it is a disturbing fact of life in exhumations that tissues and organs are often found to be absent without leave or, worse, without notation on the autopsy report.

Some pathologists are trophy collectors without permission, and without let or hindrance until caught. Witness the pathologist in Lubbock, Texas, who retained a woman's diaphragm and her breast implants after her autopsy and then, more unsettling yet, displayed them on the desk in his office. And the worst of all was my former student, Dr. Ralph Erdman, who, while in Texas, was paid for autopsies and issued reports on them when some were never performed at all.

We worked until 6:00 P.M. and finally quit in exhaustion, with Mike Baden, radiograph in hand, bent on lecturing us on the significance of Harris lines detected in long bones. Like the fungal tunneling one sees in hair from a grave site, he was telling us, in his usual learned and engaging manner, that bones show distinct transverse lines which can be mistaken as fractures when in fact they represent nutritional deficiencies that occur in bones during their developmental stages. Albert DeSalvo was known to have been subjected to the poor nutrition of institutional food during a confinement during his adolescence, the Harris lines in his long bones now demonstrating for us the time line for such a skeletal anomaly.

The next morning, we prepared to leave for our return journey to the Puritan Lawn Memorial Park. Mike Warren had his notes concerning the multiple knife wounds he had detected, both radiographically and by inspection, in DeSalvo's ribs. Those wounds certainly could have caused DeSalvo's death, and they were inflicted on him from the front, while he was standing or, more likely, while he was lying supine and unprotected.

Once again we had to struggle to extricate the coffin from the lab to Mitchell Calhoun's pickup, which was to convey it back to Massachusetts. Kevin Watts and Leo Barry had already departed with their hearse, their duties at the John Lawrence Funeral Home calling them back. Kevin did, however, leave with me the exhumation permit, which he had assisted in obtaining to authorize the DeSalvo exhumation. That permit was something of a Charlie Brown's blanket for me in case we were stopped by the police on our trip back. It certainly did not suffice as a legal document, if any was necessary, to authorize our taking the remains across a state line.

While the final cleaning-up details at the lab were under way, Mitchell Calhoun related to me an amusing anecdote concerning some young and curious children who had been watching our goings and comings from the back entrance to Jack Levisky's lab. The children had noted the coffin and, recognizing that Halloween was imminent, had put two and two together—but thankfully, in their minds that computation made three, not four. They thought the coffin had something to do with an upcoming Halloween prank.

"What's in the coffin?" they asked Mitchell.

"Oh," he replied without revealing the lie, "it's just filled with Halloween candy."

"Really?" they said as they bounced away, satisfied in the thought that they might shortly be the beneficiaries of the candy.

Hearing of this amusing encounter, I recalled our madcap escapade during the convoy to York from Massachusetts. I looked at Mitchell and remarked, "If I had only known about the candy-in-the-coffin ruse, I might have thought of using it if we had been stopped by the security forces in New York State on the way down."

Mitchell looked at me closely and said in his laconic way, "Policemen, Professor, are not children."

Albert DeSalvo's remains were placed in a sealed metal container donated by Yorktowne Caskets and loaded back into the hearse for the return transport to Massachusetts. He was reburied in Puritan Lawn Memorial Park in Peabody that Monday, with the Reverend Patricia Long presiding. I was there, along with some team members and members of the DeSalvo family. One of Albert's grandnephews placed a single rose on top of the casket before it was lowered into the ground.

When Massachusetts Attorney General Reilly learned how we had spent our weekend, he was less than pleased. In Kelly's book, she writes that he called it a "macabre stunt." But he had not been able to stop us, nor did he or Judge Young do so at the announcement of our results.

ON DECEMBER 6, 2001, AT THE NATIONAL PRESS CLUB IN WASHington, D.C., my team members reported our results. Journalists, both broadcast and print, and from all over the world, attended the event, filling the room. A platform had been decorated with a banner from The George Washington University, around which I gathered with Drs. Baden, Goldberger, and Foran. At the lectern I welcomed the media, identified the speakers who would report their findings, and gave the background on my involvement in the investigation and the studied noninvolvement of the Massachusetts authorities.

Going right to the quick, I declared, "The question we set out to answer was, Did Albert DeSalvo rape and murder Mary Sullivan?" Behind me on a large screen, Mary's face came into view, projected from a 35-millimeter slide. I described the tragedy of her murder without the specifics of her position and condition when found by her roommate. I was mindful that Diane Dodd and Casey Sherman were seated expectantly in the front seat of the room and that more details of Mary's death would just add unnecessarily to their anguish.

Mike Baden was the second man at the lectern. He described the condition of the remains and compared what he had observed to DeSalvo's confession, pointing to gross disparities in the time-of-death estimation. He also gave the audience a primer on the hyoid bone and its relation to the supposed dual strangulation, manual and ligature, by DeSalvo. Finally, he mentioned the failure to find any trauma to Mary's skull said to have been perpetrated by a well-muscled prizefighter who had confessed to battering Mary's head with his fists.

Bruce Goldberger, our toxicologist, came next on the program with an array of 35-millimeter slides that detailed the laboratory work he had done in the search for drugs or alcohol in Mary's tissues, commenting on the unusual state of preservation, which he could not explain. His test results showed no evidence of ethyl alcohol or controlled substances (illicit drugs) or even any over-the-counter drugs or any by-prescription medications in Mary's tissues. These findings eliminated the possibility that her submissive behavior, as stated by DeSalvo, was due to any external source such as drugs or alcohol—that is, if DeSalvo was to be believed on this subject.

Our final speaker provided the climax with the results of the mtDNA testing accomplished by his students under his guidance. Dr. David Foran, head of the molecular biology lab at The George Washington University, was in possession of the results of the DNA analysis, but no one other than I and one other person was privy to them. Admittedly, I had called Richard DeSalvo's son that very morning to give him the results so that his father would not be in the dark by reason of his absence from the National Press

Club briefing and would not learn first from the press what the mtDNA tests revealed.

This was the moment for David Foran to garner the justly deserved kudos that his work gave him the right to claim. I told him that he and he alone would have the floor to announce his results, in spite of the threat of reprisals I had received from others who insisted that they had a right to the results before the media. Everyone had their own claim to a right of first notification, but David Foran had the greatest of them.

As I recall his detailed slide show and his remarks, David explained that he had first looked for foreign mtDNA under Mary's fingernails, which we had retrieved at the autopsy. The fingernails can often be the repository of evidence incriminating a suspect in a forcible rape, where the rape victim scratches the assailant with her fingernails in self-defense. However, the only mtDNA Dr. Foran found on the fingernails was that of Mary herself, which was in accord with our expectations since she was found with her hands bound together.

But then, with his low-key nonchalance, David remarked and, for effect, reiterated that the various samples he had tested from Mary had revealed the presence of mtDNA from two individuals. The head hair unfortunately had not provided suitable DNA for amplification, but the pubic hair specimens told a different story. The translucent, entangled material in the pubic hair, when separated, proved to contain the mtDNA of some third person, certainly not Mary Sullivan.

The second donor of the mtDNA on Mary appeared when the underpants worn in death by Mary were examined. These mtDNA results presented David with a weighty conundrum. Two separate stains gave replicable results, indicating that the donor of the mtDNA in both stains was the same person. But—and here's the rub—the mtDNA on the underpants did not match that found on the pubic hair material. Now he had discovered two different persons' mtDNA obtained from different test samples, all of which originated with Mary Sullivan or her clothing.

Of course, at this thirty-six-year remove we had no way of knowing who prepared Mary for burial, or the source of the underpants, or whether the underpants were stained by postmortem drainage from her vagina. Yet again, a terra incognita confronted me as a consequence of an exhumation.

Will the fates never let me be to answer one mystery without putting another inscrutable one before my eyes?

Now it was the mtDNA of Richard DeSalvo, as a stand-in for his brother Albert, that was on the table, a table that David Foran was masterfully filling with suspenseful anticipation.

It was reported, rather matter-of-factly, by Dr. Foran that Richard's mtDNA sequences in the target mtDNA region were by one or more base pairs (nucleotides) different from both the underpants' mtDNA and the pubic hair's mtDNA. Without expecting his largely nonscience-literate audience to comprehend the significance of that statement, he took pains to explain that, therefore, Richard was excluded as the donor of any of those stains. Then he paused while the assembled listeners leaned forward in their chairs. No one was dozing.

Thus, he intoned as if he were standing in a pulpit, Richard must be excluded as the donor of any of the stains we tested on Mary, and so, necessarily, must his brother, Albert, be eliminated as the donor.

Now David was in his element as a lecturer qua scientist. He stated his conclusion as inescapable that no bodily fluids attributable to Albert DeSalvo had been found on Mary Sullivan's remains. On the contrary, two distinctly different persons had left their mtDNA in two separate locations, one on her underpants and the other in her pubic hair.

Moreover, it was likely that the pubic hair specimen was from her attacker, he said. To him that was certainly the simpler and most straightforward explanation for its presence. Hearing that punch line, I believed that Bishop Occam would approve, since under his rule of thumb the simpler hypothesis should be accepted when two hypotheses explain the data equally well.

What I and the National Press Club audience heard in David Foran's presentation was that it was clear beyond peradventure that his scientific findings gave no credence to Albert DeSalvo's being her attacker. This was news of a blockbuster variety. For the first time in all these thirty-six years facts had been stated that upset, upended, and unsettled the official version of Albert DeSalvo's guilt, certainly of killing Mary Sullivan, and conceding that, possibly one or more of the other Strangler victims as well.

The National Press Club program ended with Casey Sherman at the platform, urging that the search for the real Boston Strangler should and must now commence not only in earnest but for the first time. He was gratified that his conviction concerning Mary's death had been vindicated, but more to the point, he still wanted to know who killed Mary. As he put it in concluding, she died when she was only nineteen. We must remember that and her.

But the rest of the story, as Paul Harvey speaks of it, has not yet been told.

The Texas House of Representatives in 1971, while Albert DeSalvo was confined on his rape convictions in the Massachusetts State Penitentiary at Walpole, topped off the DeSalvo–as–Boston Strangler legend by unanimously approving a resolution, submitted by Representative Moore, focusing on DeSalvo's accomplishments. But query? What were these so-called "accomplishments" for which he was commended by this resolution?

After six "Whereas" clauses, the Texas House of Representatives commended "Albert DeSalvo on his outstanding career of public service." But yet again, what "public service"? one might rightly ask. In explanation, according to the fourth "Whereas," DeSalvo had "been officially recognized by Massachusetts for his noted activities and unconventional techniques."

And, pray tell, what "noted activities and unconventional techniques" were these? They were said to involve "population control and applied psychology." Moreover, DeSalvo was applauded for enabling "the weak and lonely . . . to achieve . . . and maintain a new degree of concern for public safety." When it was learned who Albert DeSalvo truly was, the unanimously approved resolution was unceremoniously and unanimously withdrawn.

Now, that is the rest of the story.

6. THE UNHEARD VOICES
OF THE DEAD

Behind the corpse in the reservoir, behind the ghost on the links,
Behind the lady who dances and the man who madly drinks,
Under the look of fatigue, the attack of migraine and the sigh
There is always another story, there is more than meets the eye.

W. H. AUDEN, "AT LAST THE SECRET IS OUT"

The dead can have no voice, I have sometimes learned to my regret, unless they are fairly given the chance to speak. I tremble at the thought and the knowledge that the dead can be denied this "last right" by the cold, calculated, or curious attitudes of those who control the decision to grant such a last rite.

The reasons the living present in opposition to hearing the voice of the dead are multifarious, some masking the true motives that the objectors dare not articulate openly. It is one thing to be forthright and admit to an unshakable resolve against disturbing the dead, although, being of a practical mind, I have never understood how people who are dead can be disturbed by an exhumation, except possibly through the emotional attachments of the living to the dead.

Of course, I recognize and respect the religious views of those who believe that the dead should lie where they have been buried, untouched and inviolable for all eternity, or, for some, until the Second Coming of Christ. As part of my considering conducting an exhumation, I always inquire about the religious views of those whose permissions my desiderata for an exhumation stipulate that I should obtain, most particularly the owners of the cemetery where the deceased is interred.

Once, in my work at Harewood, the home of Samuel Washington in West Virginia, I was reliably informed by the descendants of African-American slaves who had worked at this colonial mansion that it would violate their sensitivities if any African-American bones that might, by happenstance, be unearthed were to be wrapped for transport or otherwise in newspaper. I did not delve into the reasons for such a stipulation, nor did they offer any to me. Compliance was a softer touch than contention.

Aside from faith-based objections to an exhumation, there are those who view the matter of an exhumation as "a dirty deed" and not a way to tell a "gallous [gallant] story," as Pegeen Mike said, in another connection, in Synge's great play *The Playboy of the Western World*. To such persons, the subject is grimly ghoulish, akin to the actions of the resurrectionists of "Merry Old England," who went about in the dark of night prowling grave sites for the newly buried so as to profit from trafficking in their cadavers or from the valuables that might have been buried with them. Even today, graves have been found to be vandalized, as happened in the early-nineteenth-century cemetery known as the Blackburn Cemetery, located just off the Natchez Trace Highway in Tennessee. The persons who had been buried there had died contemporaneously with Meriwether Lewis, whose grave is nearby but, happily, unvandalized.

Sadly, those who deserve the stewardship of their graves to be strictly regulated have had their graves vandalized along with those, as in the Blackburn Cemetery, who lie unprotected and beyond the watchful eye of either passersby or groundskeepers. Gouverneur Morris, whom his biographer Richard Brookhiser terms "The Rake Who Wrote the Constitution," is unfortunately one of those, interred as he was in a vaulted tomb and cast-iron casket that have been vandalized. Morris died in 1816 and was buried at St. Ann's Episcopal Church, now on St. Ann's Avenue in the South Bronx, New York City. In 1986, anthropologist Peggy Caldwell-Ott received a telephone call asking her to inspect the vandalized casket of Gouverneur Morris.

Her report indicates the presence in Morris's casket of substantial parts of two skeletons, one a male thirty-five to forty years old and the other a female. Since Morris died at sixty-four years of age, neither of the skeletons in his casket could be associated with him. So the question is: Where are the remains of Gouverneur Morris? No one with authority to consent seems to be in the slightest degree interested in finding him. I have been informed that the church has nixed going forward. More's the pity that a man of such stature in the history of this country should not have a voice in the retrieval of his remains.

Another basis for a rock-ribbed rejection of an exhumation stems from the desire for closure, especially on the part of relatives of the deceased. The burial of the remains, it is thought, puts a final imprint on the death, the pain from which is only renewed and revived by an exhumation. Particularly tragic deaths are of that character, such as those denominated as deaths occurring by asphyxiation during an autoerotic episode. Where a medical examiner chooses to term such a death a suicide, that may be the impetus for the family to seek a redetermination to an accidental death via an exhumation. But often that impulse is squelched by the greater good, or so it is perceived to be, of not reliving the death, with family members who are often the first to find the body and the sexual accoutrements used in the deceased's final fatal sexual venture.

Sometimes the family of the deceased, whether the immediate family or descendants many generations removed, is torn between permitting an exhumation and rejecting the idea for fear of the publicity and notoriety that it might bring. This concern dogged the descendants of Samuel Washington in my Harewood underground explorations in 1999. Their apprehensions over giving public voice to my efforts were twofold. First, they had no fondness for the media, broadcast or print. As John Washington, a direct descendant in the male line from Samuel, put it to me when I presented him with the opportunity to have a British television team on-site during our two weeks of activities, "Tell them you're working in California." As a consequence of that fixed position of the Washington family, I proceeded without the presence of the press or any contact with the press, either local or more widespread.

Such a reluctance to indulge the media also played a role in the preparatory stages of my Jesse James exhumation. Betty Barr, a known descendant
of Jesse James, was at first horrified at the thought of a media event or cir-
cus, as some label it pejoratively, occurring at her ancestor's grave site. She,
unlike the Washington family, was persuaded to allow the exhumation on
condition that the media would be at all times rigidly controlled by me and
not allowed, as almost happened with a *Los Angeles Times* reporter in 1989
at the Packer victims site, to roam at will at the Mount Olivet Cemetery or
to besiege her with interview requests after the fashion of the paparazzi.

Naturally, the dead person may in life make his or her desires known as
to whether a privately sponsored exhumation would be acceptable to him
or her. That can best and most formally be accomplished through a will
declaring the testator's wishes on the subject of his or her burial or even
exhumation, which is not, as it happens, as common a clause in a will as
the burial arrangements are.

Headstone inscriptions should not be relied upon to govern the expression of the decedent's wishes. William Shakespeare is said to be (there
are those who claim this is not his genuine burial place) buried in a chan-
cel plot in Holy Trinity Church in Stratford-on-Avon, Warwickshire. The
monument on his supposed grave bears an inscription that reads:

Good Friend, for Jesus' sake forbear
To dig the dust enclosed here;
Blessed be the man that spares these stones,
And curst be he that moves my bones.

That inscription is a definite and clear-cut statement that Shakespeare's
bones (dust) should not be subjected to an exhumation. And yet is it, in
truth, in the words of Shakespeare or of some poseur, possibly even Sir
Francis Bacon, who is often thought to be the author of one or more of
the plays for which Shakespeare is given credit?

The inscription cut into the headstone of Ireland's famed literary figure William Butler Yeats in the Drumcliff churchyard in County Sligo,
Ireland, is unquestionably in his words. The inscription, taken from the
last verse of his poem "Under Ben Bulben," states:

Cast a cold eye
On life, on death.
Horseman, pass by!

But even though it's admittedly in his words, what is its meaning? Is it an injunction, like Shakespeare's, against tampering with his remains? Or is the horseman referred to the more traditional "grim reaper" whom he wishes would "pass by"? In short, the living are admonished not to leave their wishes in regard to an exhumation up for linguistic grabs on a monument or a headstone. Why leave room for persnickety lawyers?

On occasion, even without an objection from any authoritative source, my exhumation has fallen on fallow grounds, insufficient for the answer to my quest to bloom. For example, I was invited by Angela Shearer, a master's candidate at The George Washington University, to initiate a search for the unmarked grave of Samuel Washington, brother of George, but despite massive digging for two weeks in 1999 we were unable to identify remains that had been his. Other remains of his family were discovered, but not his, the object of our search.

Planting evidence at a dig site is not an unknown occurrence. At the Samuel Washington dig, I was directed by our backhoe operator, Bud Ridenour, to an artifact that had been upturned by his backhoe. It turned out to be a decorative antique coffin handle suggestive of the presence of a coffin or, better yet, human remains. Neither was found after intense scrutiny and screening.

At a later date, Bud, with a beaming smile of innocence creasing his face, informed me that when he saw that my team was in the doldrums over our failure to find human bones, he thought the planting of the coffin handle would be a welcome pick-me-up for us. In the vernacular, upon hearing this confession I could have croaked, for I had been taken in and was now confronted with the obligation to explain the presence of this spurious artifact that we had logged in quite unsuspectingly.

In Gettysburg at the Daniel Lady Farm, which had been the site of a Civil War Confederate field hospital during the Battle of Gettysburg in July 1863, I was persuaded to call on a cadaver-sniffing dog, Eagle, whose handler, Sandy Anderson, came with distinguished credentials. To be a mite more scientific in our search, I also supervised the use of metal detectors and ground-penetrating radar, courtesy of geophysicist Greg Baker, a professor at the State University of New York at Buffalo A bone dowser, without my knowledge or permission, and out of my presence, also attempted to locate the long-forgotten graves of long-dead Confederate soldiers believed by the Gettysburg Battlefield Preservation Association, the owner of the Daniel Lady Farm, to have been buried there and not recovered in the years following the battle.

After two weeks in the blazing sun, on a June day in 2003, a clavicle, and only a clavicle, was unearthed by Eagle in the ground near the barn. Of course we had already found a motley grouping of artifacts deposited at the site from 1863 and before to the present. We packaged such artifacts separately, with identifying numbers and dates and locations of when and where they were found. In doing so we were only playing at being archaeologists, since our driving interest was in finding human bones.

The barn was a likely place to find human bones, for it had served during the battle as a surgery for the enlisted men. Although we labored mightily, with a steadfast and eager team of volunteers and a number of determined and muscled diggers, we did not find anything of human origin other than this clavicle. It was later identified by anthropologist William Bass, Ph.D., as a male clavicle in remarkably and possibly suspiciously well-preserved condition after 150 years in the ground.

With the experience of Bud Ridenour still in mind, I considered the possibility that the dog's finding the clavicle might also have been a sleight-of-hand, in this case a sleight-of-dog. As it turned out, I learned that a Michigan grand jury had been investigating a claim in another matter that Eagle, the bone-sniffing dog, in league with her handler, Sandy Anderson, had hit upon bones that were said to have been planted to prove her mettle or for other reasons. The disposition of that investigation is, at this time, under wraps.

At other times, I have been ready to investigate a historical controversy but persons, not extraneous circumstances, hindered me from coming within reach of the prize. One such case involved an infamous double murder from the nineteenth century that has given rise to undimmed controversy ever since its occurrence.

Lizzie Borden took an axe
And gave her mother forty whacks;
And when she saw what she had done,
She gave her father forty-one.

AUTHOR UNKNOWN

It was on the morning of August 4, 1892, that the mutilated bodies of Andrew Borden, seventy, and his wife, Abby Borden, sixty-five, were discovered in their two-story home in Fall River, Massachusetts. Andrew's corpse, found first, lay faceup on the living room divan, his bloody head and face brutally cut by eleven blows from a sharp implement. Soon Abby's body was discovered on the second floor in the guest room that fronted the street. She, too, had been slain with a sharp weapon, inflicting upon her skull and upper back some eighteen to twenty blows.

Andrew's thirty-two-year-old daughter, Lizzie, a matron and Sunday-school teacher who found his body, was arrested and tried, but she was found not guilty by a jury of twelve men after a trial founded entirely on circumstantial evidence, including scientific opinions bordering on, in today's world of forensic science, quackery. She lived out the rest of her life as a woman of means in a home called Maplecroft in Fall River, taking frequent excursions to the far less mundane life of New York City. She died in June 1927, just eight days before Emma, her older, also spinster, sister and another suspect in the double murders. She is buried at Oak Grove Cemetery in Fall River, with a marker in the Borden family plot simply stating her name as "Lizbeth."

In spite of Lizzie's acquittal, whether or not she was actually the killer has remained a matter of much disagreement. Scholars and historians have lined up on both sides, including prominent literary figures. Agnes de Mille's ballet *Fall River Legend* portrays Lizzie as guilty of the murders due to the stuffiness of the small-town atmosphere of Fall River, giving rise to her claustrophic mind-set, leading to murder as an outlet.

Abby and Andrew were buried in Oak Grove Cemetery. On the day of the funeral for them, only a few days after the murders, the burial was halted. At the grave site, the police informed the mourners that a medical doctor wished to conduct an autopsy, so the bodies were removed to permit the autopsy as well as the disarticulation of the heads. The headless bodies were then buried.

The heads of Abby and Andrew were macerated (defleshed) to be made available as exhibits at Lizzie's trial. Only one skull, that of Andrew, was put into evidence as a demonstrative exhibit at the trial. A Boston physician, Dr. Frank Draper, demonstrated that the blade of what was then called the "handleless hatchet" obtained from the Borden home fit the cut marks in Andrew's skull. He held the skull in his hand and put the hatchet blade into the cut marks, declaring that the hatchet was likely to be the fatal weapon.

Dr. Draper's testimony of a fit, when viewed in the clarifying light of hindsight, does not measure up. In the world of modern forensic science, his testimony was rubbish. Any tool mark examiner knows that the microscopic markings left by a hatchet blade in the softer surface of the skull are the only basis of individualizing the hatchet blade to the cut marks in the skull structure.

It was my plan to use up-to-date tool mark analysis via a microscope to determine whether this hatchet, now in possession of the Fall River Historical Society, after having resided for many years in a hip tub in the attic of Lizzie's lawyer, was indeed the murder weapon. If the cut marks in the skulls were shown not to have been inflicted by the hatchet produced at Lizzie's trial, that would be a significant result. While this finding would not exonerate her entirely, it would reveal a central weakness in the prosecution's case against her, which was heavily reliant upon the handleless hatchet's being the murder weapon.

But where were the skulls of Abby and Andrew? They disappeared after Lizzie's trial without any documentation as to their whereabouts. They continue to be among the missing. But just on the off chance that they might someday come to light, I sought permission to examine and to cast the blade of the handleless hatchet.

Michael Martins, curator at the Fall River Historical Society, graciously granted me the permission requested. With a television crew in tow, I measured and photographed the hatchet, and Mikrosil-cast the blade of the "handleless hatchet," which had been in the custody of the Society since 1968. I forwarded the casts to a forensic laboratory in Oregon, where they were examined and photomicrographed under a scanning electron microscope, all of which was intended to reveal the condition of the blade under high magnification.

The photomicrographs did not lie! The hatchet blade was in just-off-the-shelf condition. It bore the distinctive tool mark impressions of a recent manufacture. This news was good cause for elation.

Contingent on the present condition of the skulls, if this hatchet was indeed the murder weapon, the manufacturing imperfections detected in the hatchet blade should also be in the cut marks in the skulls. But first I had to locate the skulls. The hatchet alone would not tell the tale. The location of the skulls of Abby and Andrew was now a priority item for me, for a matching to the hatchet could not be conducted without those skulls, or, at least, one of them. That is elemental forensic science, my dear reader.

Before leaving the Fall River Historical Society's Victorian mansion, I was shown two black-and-white pictures, said to be of the skulls of Abby and Andrew. It was intimated that the pictures might suffice in lieu of the actual skulls. I brushed off that suggestion without a moment's hesitation. It was, I said, the three-dimensional skulls and only those that would serve my purposes, for once again it was imperative for me to cast the cut marks to compare them to the imperfections on the hatchet blade. Since when, I mused rhetorically, can one cast cut marks in a two-dimensional picture, even if the cut marks are plainly visible?

But the pictures themselves were disturbing, not only because they showed the heavily tortured and fractured skulls of Abby and Andrew but because each picture was assigned to one or the other of the two victims. However, there just were not enough cephalic markers in view to be able definitively to say which skull belonged to whom, Abby or Andrew. Sexing of a skull is a difficult and uncertain task in any case, but without clearly demarked indicia such as the occipital protuberance (crossing the rear of the skull laterally), the mastoid process (a protective cover for the ears), and the supraorbital ridge (above the eyebrows on the frontal bone), at a minimum the task of sexing would be most problematic.

I don't know whether Mr. Martins took my views to heart or whether he left the pictures displayed, now knowing that the pictures were quite possibly mislabeled. I was to learn in my later investigation into the death of

Jesse James that pictures and such on the wall are attractive to the tourist traffic, which can be claimed as an excuse for retaining an erroneous depiction.

My search for the skulls took me, quite plausibly, back to the Oak Grove Cemetery, where in the handwritten notations recording the burials of Abby and Andrew there were nearly illegible and undecipherable handwritten entries above the notation of the date of the original burials. It looked to me to have been made after the first entry for Andrew and Abby. I reasoned that these might signify the later burial of the skulls.

To test this hypothesis, I asked and received permission from the cemetery to use ground-penetrating radar to scan the Borden graves. The scans were conducted on two different dates, one in warm weather by New York University geophysicist Jim Mellett, Ph.D., and another in the more electromagnetically responsive hard cold ground. The printouts of the second GPR runs, conducted by the same Stan Smith who came to my aid in the Alfred Packer explorations, satisfied me that there had been two different burials: first, the headless remains, and later, above them, another burial. I surmised that these subsurface anomalies signaled the presence of the skulls above the coffins of Abby and Andrew, although I would need an exhumation to confirm that theory.

On the occasion of the second GPR scanning, at about the time I was in readiness to depart from the cemetery, a television crew arrived carting a rocking chair. My curiosity whetted, I asked what they were about. I was informed that a psychic—she was pointed out to me—was going to see what vibrations she could detect while rocking over the graves of Andrew and Abby. Following that search for "vibes," she would move to Lizzie's grave to learn what her subconscious might tell her about the vibrations there. I decided to postpone my departure. This experiment was worth watching, even though I was thoroughly skeptical that anything but a good laugh would come of it. Little did I know.

My cameraman, Jim Kendrick, on loan for the moment from The George Washington University's Medical Center photography department, and I scrambled off the Borden plot and were shushed into silence by the television people. As we watched, the psychic was seated in the rocking chair and was presented with a heavy ax, which she cradled in her arms.

Somehow the singsong children's quatrain about Lizzie's taking an "ax" and giving Abby and Andrew so many whacks with it had captured the imagination of the television people. Of course, anyone even remotely familiar with the historical record concerning the death of the Bordens and the trial of Lizzie knows that the Bordens were hatcheted to death and were not the victims of an ax-wielding murderer. I suppose if a chain saw would fit the rhyme there might be some who would make Lizzie out to be a "chain-saw massacre-ess." That might suit those who have dubbed her as having the face and the bearing of a "concentration camp madam."

As Jim and I watched the psychic quietly in action in the rocking chair, I began to be concerned that her moving from a gentle to a more daring pace might cause her to lose her balance. I whispered to Jim that if she was not more careful the chair would be propelled backward, causing her to make a most red-faced exit.

Just as I stated my prediction of danger, psychic though I am not, she caught herself before she went over backward and, in overcompensating, careened forward, falling from the rocking chair to the ground, the ax still in her arms. As she fell, the cameras rolling, she screamed, "I'm cut. I'm dying."

We all ran to her rescue, thinking that the ax blade had laid her low. No, far from being hurt, she was unscathed; only her coat had been torn by the blade of the ax.

As things returned to normal for this psychic of the abnormal, Jim Kendrick, good old whimsical Jim, turned to me and said, "Some psychic she is. She could not even see it coming."

Lighter moments like these tend to leaven the exhumation enterprise.

My next step was to obtain the consent of the living Borden relatives to an exhumation. But the officials at the Fall River Historical Society, where the known relatives were recorded, refused to provide me with the names and addresses of any of the relatives, claiming that they had an obligation of confidentiality to them. I was thus stymied in trying to reach anyone in authority to grant me the requisite permission.

And Lady Luck did not shine on me in that continuing effort. After examining the cemetery records, I asked a cemetery staff member about a suitable place for lunch. A helpful Hannah recommended the Lizzie Borden Café in downtown Fall River. Then someone notified the local press of my investigations and the newspaper editor joined me, uninvited, at the café. The next day, the headline in the local newspaper proclaimed my intent to exhume Andrew and Abby Borden's remains. It wasn't long before I heard from people claiming to be the Bordens' relatives, and the missives were sorry and scandalous, even slanderous, statements of vehement and vituperative opposition to any exhumations of the Bordens. Never had I been so pummeled in my investigations.

I was accused of scheming all sorts of indiscretions concerning the Bordens' remains. The letters, of the chain variety, denounced me for trampling on the rights of the dead and for being no more than a publicity seeker who hoped to upset the common conviction that Lizzie was guilty of the murders and had beaten the rap.

A most quixotic feature of all of these letters did not escape my attention. The relatives, claiming to be related to Lizzie through Andrew, her father, were solidly of a mind that Lizzie was guilty—as guilty as sin, as the saying goes—even though she had been acquitted. These purported relatives feared that my scientific analyses would prove them wrong and confirm the jury's verdict of acquittal. This was to be the first and only time in all of my investigative experience that relatives were happier in the belief that their relation was guilty of heinous crimes even though, de facto, through a jury's verdict there had been an acquittal. Thankfully, paradoxes of this most peculiar and inexplicable sort do not appear on my investigative plate with regularity.

But another detail was of greater moment. I noticed that all of these poison-pen letter writers claimed to be descendants of Andrew Borden, despite the fact that Lizzie and her sister, Emma, had been childless spinsters and had no brothers. So the claims to being in a direct line of descent from Andrew could not have been true. Moreover, none of the letter writers claimed kinship to Abby Borden, Lizzie's stepmother. Thus, the remains of Abby were still fair game for an uncontested, at least by these letter writers, exhumation court order. However, it is not my policy to precipitate an adversarial contest in a court of law, with my opposition being those claiming kinship to the deceased. I was certain that the objecting letter writers, even without any provable standing as to Abby, would intervene and protest with vigor and vitriol my seeking a court order for an exhumation.

I decided on a flanking maneuver that might still produce scientific results without an exhumation and give new meaning to Lizzie's guilt or innocence. My visits to the Fall River Historical Society had alerted me to its having custody of a number of artifacts from the Borden killings other than the hatchet. Hairs said to be human and to have been found on the hatchet blade were on display at the society's headquarters, as was a stained coverlet—stained possibly by blood—from the upstairs bedroom where Abby had been killed. The hairs and the stained coverlet were ideal subjects for analysis, the hairs by the nondestructive means of microscopic testing and the coverlet by destructive testing through either DNA profiling or more long-standing physiological stain analysis (commonly known as serological testing).

I wrote the Fall River Historical Society to request permission to conduct testing on the hair and bloodstains. As regards the coverlet, I recommended taking just a few fibers from an area well-removed from the main stain, but even my conservative approach baked no loaves.

The response from a Society official was a complete and unequivocal rejection of my requests. The nature of the reply was so unnecessarily protective of the material in question that it got my Irish up. I then wrote back, with words of unconcealed pique, pointing out that the Fall River Historical Society was giving more dignity to the artifacts of the Borden killings than the Vatican had done with respect to the purported face of Christ on the Shroud of Turin, which had been subjected to destructive carbon-14 dating. That ended the correspondence between us, as well as my attempts to have science play a role in solving the many mysteries of this case.

But I was granted leave to play one more card in the game of Lizzie's guilt or innocence. Even though this card was not a trump card, it gave me much personal satisfaction. Jules Ryckebusch, a teacher at the Bristol Community College in Fall River, decided not to let the hundredth anniversary of the Bordens' unsolved killings pass unnoticed. He therefore assembled a group of key players in the Borden drama for a convocation at the Bristol Community College from August 3 to 5, 1992.

I was invited to present my work to the crowded auditorium of Lizzie-ites. The conference schedule called for me to deliver my remarks in a

35-millimeter-slide format on August 4, 1992, in the morning. That scheduling was unnerving not only because it coincided with my granddaughter, Shannon's, birthday but because I would be at the podium speaking at precisely the moment, by day and hour, when one hundred years before, the hue and cry went out that the Bordens had been murdered. However, I survived the moment, even though the Bordens hadn't.

Since my involvement in investigating the Borden murder, I have waited anxiously for notice of the release of the notebooks of Lizzie's trial attorney, Andrew Jennings. Even though both Lizzie and her attorney are long dead, those notebooks, containing confidential communications from Lizzie, might end the debate over Lizzie's role in the murders. Apparently, those notebooks are in the custody of Andrew Jennings's superseding law firm, which is still in practice.

However, the United States Supreme Court, in litigation involving an attorney's notes during an interview with Vincent Foster of the Clinton White House, who later committed suicide, has refused to require the production of those notes. Consequently, the confidentiality of the attorney-client relation extends beyond the grave, at least in those states that rigidly follow the common law of posthumous confidentiality. Thus, if new information is to emerge, it most likely will be generated by science, either with or without an exhumation. I stand ready and await the call to dig at the Oak Grove Cemetery or to test the artifacts at the Fall River Historical Society.

However, of more urgent and immediate concern to me is an exhumation of the remains of Meriwether Lewis, one of our national heroes, from his grave lying under a broken-shaft monument in the high hills of middle Tennessee some seventy or so miles south of Nashville.

The garlands wither on your brow;
*   Then boast no more your mighty deeds;*
Upon Death's purple altar now
*   See where the victor-victim bleeds.*

JAMES SHIRLEY, "A DIRGE"

Some deaths are so electrified as legend that the truth may be known, if at all, only with the modern methods of science, and where the legend outruns the truth, then, as with Jesse James, an exhumation may well be in order so that the truth will surpass and even dispel the legend. Meriwether Lewis's death is a prime example of the necessity to right the imbalance between far-fetched legend and inestimable truth.

My involvement in the issues surrounding the death of Meriwether Lewis occurred quite serendipitously in the early 1990s when my wife, Barbara, and I were returning home from a Florida family wedding. We had been traveling by car on the Natchez Trace Parkway from Tupelo, Mississippi, northward toward Nashville. At my wife's urgent insistence we stopped many, many times at the sites of historical markers along the roadway, she being a dedicated and quite learned amateur historian.

One of our intermittent stops was at the "Pioneer Cemetery," where a rustic cabin stood uncontrolled and open to all passersby, and there were surely few of those in this somber, lonely woodland location. A short walk away, but in sight of the cabin, was a broken-shaft monument. As we stood looking up at this graffiti-desecrated monument, my wife remarked quite noncommittally that this was a death I should investigate with the instruments of science. We sat leaning our backs against the base of the monument, with Jefferson's words to Meriwether Lewis inscribed above us: "His courage was undaunted: His firmness and perseverance yielded to nothing but impossibilities . . ."

Barbara gave me a lesson in basic facts underlying the controverted death of Meriwether Lewis. Her informed description added to my sum total of the man as I knew him from the abbreviated *Journals of Lewis and Clark* by John Bakeless. I still have that much-paged paper edition that Barbara read to our children in the late 1960s as we journeyed via a VW bus across the country, following as best we could on roads near the land and water route of the Lewis and Clark Expedition, known as the Corps of Discovery. At that time, like so many Americans, I knew of the expedition but not what happened afterward, certainly not of the deaths of its leaders, Lewis and Clark. My wife's detailing the facts of Meriwether Lewis's death set me on my own journey of exploration, seeking the answers to the tragedy and the dilemma of Meriwether Lewis's death. That journey has continued to this day, and I am resigned to continue it unflaggingly in the future.

An examination of the circumstances and documents associated with

Meriwether Lewis's death provides clear evidence that no one has proof beyond a reasonable doubt as to the actual manner of his death, whether it resulted from suicide or homicide. The only evidence that remains to clarify this issue are the bones of Lewis, buried some seventy miles south of Nashville, Tennessee, an outfielder's throw from the Natchez Trace. A broken-shaft monument (symbolizing a life cut off in its prime) marks the place where his remains rest.

For almost three years, from 1803 into 1806, Meriwether Lewis and William Clark prepared for and endured the incalculable challenges of a water and land course across the North American continent. They led a band of thirty volunteers (depending on whether you count Clark's slave, York; Charbonneau's Shoshone wife, Sacajawea; and Newman and Reed, who were discharged en route) from St. Louis, Missouri, all the way to the Pacific Ocean and back. They were the equivalent of Australia's Robert O'Hare Burke and William John Wills, who in 1861 crossed Australia on foot from south to north, except that Burke and Wills died of starvation at Cooper's Creek, tragically within an easy jaunt to the terminus of their journey.

President Jefferson, in written instructions to Lewis dated June 20, 1803, stated that "the object of your mission is to explore the Missouri river, & such principal streams of it, as . . . may offer the most direct & practicable water communication across this continent, for the purposes of commerce." Subsidiary objectives were to gain knowledge of the inhabitants they might meet as well as the geology, the flora and the fauna, the astronomical sightings from place to place, the topography of the land, and much more. Jefferson's instructions left it to Lewis to dot every scientific *i* and to cross every conceivable *t*. Lewis's observations were directed to be recorded "with great pains & accuracy" in a journal "made at leisure times."

Above all else, Lewis and Lewis alone, a man not yet thirty years of age, was entrusted with being in sole command of the journey, its men, and its day-to-day functioning. Most important, he was admonished by Jefferson to "err on the side of your safety, & bring back your party safe." All this Jefferson asked of a man he would describe in disparaging terms some four years later in his *Memoir of Meriwether Lewis,* dated, most strangely, August 18, 1813—August 18 being Meriwether Lewis's birthday. As Jefferson put it, Lewis had been "from early life . . . subject to hypochondriac affections" that Jefferson had observed during Lewis's time in Washington as his secretary as "depressions of mind."

Which Jefferson are we to believe, the Jefferson who on June 20, 1803, put his entire trust in the sanity and stability of Lewis, or the Jefferson of August 18, 1813, a Jefferson who had known in 1803 and before that Lewis was a depressed hypochondriac? Would such a man, so afflicted with a serious mental infirmity, be put in command of the safety and well-being of thirty-odd explorers who held the destiny of the country's immediate westward expansion in their hands and their pirogues?

These men managed to accomplish all that was asked of them, but an equally remarkable feat was that they succeeded in this perilous journey without the loss of a single man to the hazards and dangers of the journey. Sergeant Floyd's death in Iowa was probably due to natural causes, possibly septicemia from a ruptured appendix.

Only three years later, on October 11, 1809, Lewis would be dead, at the age of thirty-five. At the time, he was governor of the Louisiana Territory, the vast lands purchased for a pittance from Napoleon. Some historians feel certain that his death resulted from a suicide. Others are unsure. Still others call it a homicide. An examination of the considerable array of documents surrounding and commenting upon the incident is proof positive that the event is steeped in mystery.

Lewis had embarked on a journey to Washington, D.C., with his as-yet-unpublished journals to plead for the payment of debts he had incurred and which the government, he maintained, "stiffed" him in failing to reimburse. Worse yet, the government had strongly suggested that his dealing with the Indians was in conflict with his role as governor.

He set out from St. Louis in good and unwavering spirits on September 4 with his servant, Pernier (sometimes called Pernia), but along the way he became ill. He rested at Fort Pickering in Tennessee (near present-day Memphis), leaving two trunks of his belongings there, then left on September 29, accompanied by Major James Neelly (sometimes Neely), the Indian agent to the Chickasaw, along with Neelly's servant, Tom, and possibly an Indian as well. They ventured east toward Nashville, veering in a southeastern direction toward the Natchez Trace. While en route, Lewis had a brief recurrence of his illness, the nature of which is not described.

On October 8, they arrived at the Indian footpath known as the Natchez Trace, where two of their horses wandered off during the night. Neelly volunteered to remain behind to find them, while Lewis traveled on with Pernier and Neelly's slave to find a place for the night. He arrived on October 10 at Grinder's Stand, a traveler's way station on the trail, where Mrs. Robert Grinder (sometimes known as Griner), in the absence of her husband, offered him lodging. That night proved to be Lewis's last.

It is rare to find a historian who has written with a skeptical eye about the claim that the party's two horses simply wandered off. It is passing strange that precautions, like hobbling or tethering of the horses, were not taken to prevent them from getting loose, especially by Neelly, the Indian agent, who must have known that horse stealing by the Indians and then exchanging the stolen horses with the settlers for provisions was a common occurrence. Is the random straying of horses just cause for casting a suspicious eye on him in connection with Lewis's demise? Given what else we know, his oversight with the horses remains suspicious, especially since his separating himself from Governor Lewis, who was supposed to be in his care, was most strange indeed.

The accounts of the cause and manner of Lewis's death vary in vital details, since the facts surrounding his death are meager and conflicting, with no eyewitness versions extant. The majority of historians and commentators on the death of Meriwether Lewis have drawn upon one or more of three sources for their interpretation of the facts precipitating it.

The first two reports (those of Neelly and Wilson) are reliant upon the hearsay of Mrs. Grinder's recollections. The third, that of Major Russell, is last in terms of its appearance and its reliability, for the informant upon whom Major Russell founds his description of Meriwether Lewis's death is not disclosed. In order of their chronology, the first of these reports was that of Lewis's traveling companion, Major James Neelly, in a letter to Thomas Jefferson, dated October 18, 1809. The second was that of Lewis's ornithologist friend, Alexander Wilson, in a letter to another ornithologist, Alexander Lawson, of May 28, 1811. And the third has been described by historian Donald Jackson as a "statement" from Major Gilbert C. Russell, an army friend of Lewis and a former officer-in-charge at Fort Pickering, Tennessee, at the time of Lewis's sojourn there. This statement is said to be dated November 26, 1811.

All three recitals agree that Meriwether Lewis suffered two bullet wounds from pistols. All three are in accord that no one witnessed the firing of the pistols inflicting these wounds. The fact that two pistols were fired during the night of Lewis's stay at the lodgings at Grinder's Stand was the version Mrs. Grinder gave in declaring what she had heard but not seen.

Major Russell, and only Major Russell, said that Lewis was "found about day light, by one of the servants, busily engaged in cuting [sic] himself from head to foot" with "his razors" while "sitting up in his bed" in his lodging at Grinder's Stand. Other commentators, reconstructing the event after the lapse of many years, have spoken of Lewis's bullet wounds as being aggravated by "a cut throat," ostensibly from his razors.

Aside from the wounds from the razors, the two bullet wounds are variously and often conflictingly related in each of these three reports. The Russell claim of Lewis's cutting himself with "razors" is well-nigh as inconceivable as declaring the "Giants' Causeway" off the eastern coast of Ireland to be a Roman-made version of the Appian Way. There are some feats we humans just cannot perform, such as cutting ourselves "from head to foot" after having been shot twice in vital bodily places by a .69-caliber handgun or guns. Not even the bold and the brave Meriwether Lewis could defy the laws of physiology by doing so.

Mrs. Grinder told Major Neelly that Lewis had been pacing about restlessly, acting "like a lawyer," and that at about 3:00 A.M., she heard two pistol shots "in the Governors room." She awakened the servants, but it was too late to save him. They found Lewis conscious, but dying from a wound to his head from one pistol and another to his chest with the other. He supposedly said to Pernier, "I have done the business my good servant give me some water." He lived only a short time after that, and Neelly buried him there, as "decently" as he could.

The suicide theorists find no ambiguity in this deathbed declaration, confidently asserting that it constitutes a firm confession of Lewis's suicidal intent. The "business" did not, to those partial to the suicide view, refer to Lewis's transcontinental journey, nor to his days in St. Louis as governor, but to, and only to, his suicidal act. Yet this entirely ignores the fact that Mrs. Grinder did not speak of Lewis's having overindulged or even tippled at all in alcohol that night.

Alcohol, according to a statistical truism, is a most frequent accompaniment to suicide, as are other central nervous system depressants in today's world. Alcohol consumption on the night preceding Lewis's death would be expected if Lewis were the alcoholic some historians claim he was, as well as if he were bent on suicide.

Alexander Wilson gave a markedly contrasting view of Lewis's death, which he had gleaned from his meeting with Mrs. Grinder. She said that Lewis had been agitated during the night, but earlier had conversed with her in a calm manner. At some point after he had retired, she heard a pistol go off in the governor's lodgings, followed by a thump on the floor and an exclamation: "Oh Lord!" Then she heard another pistol shot. Shortly thereafter, she heard Lewis scratch at her door while saying, "Oh madam! Give me some water and heal my wounds."

She watched through a chink in the door as Lewis groped his way to a tree, then back to the house, where he tried in vain to get water with a gourd from a bucket. It was two hours before she had the courage to check on him. He was still conscious enough to show her where the bullet had entered his side. She could see for herself that a portion of his brain was exposed from a gunshot wound, "without having bled much." He begged her or the servants to take his rifle and finish him off, offering them all the money he had, for he was "so strong, so hard to die." They would not do it. His death followed two hours later, just as Major Neelly arrived with the lost horses.

The third report, on November 26, 1911, more than a year after Lewis's death, from Major Gilbert Russell displayed a flourish of literary license and lack of attribution when he claimed that Lewis killed himself in a fit of paranoia over imagined enemies. He "discharged one [pistol] against his forehead without much effect—the ball not penetrating the skull but only making a furrow over it. He then discharged the other against his breast where the ball entered and passing downward thro' his body came out low near his backbone."

Despite the insistence by these three reports that Lewis died by his own hand, no one actually witnessed the shooting. In addition, the motive for his committing suicide is, at best, inconsistent and, at worst, completely lacking. The accounts of his deathbed declarations are ambiguous, a testament to the failed recollections of the reporters or their embellishment of the facts. After learning all of this, I decided that Lewis's death was surrounded in sufficient controversy to warrant a scientific examination of his remains.

I offer one other consideration as well to support the mysteriousness of the death event: Mrs. Grinder's refusal to go to his aid immediately, as Wilson's account has it, makes little sense, given her daily life in the wild, with dangers constantly on all sides. It is all too possible that she was hiding from those who had come to do, and had done, Lewis harm in conjunction with a robbery. Some historians are convinced from her statement that he was assassinated even—with Pernier's complicity.

Mrs. Grinder's behavior, as she reported it to Wilson, takes on greater meaning in view of the fact that the Natchez Trace Indian Trail where Lewis died, was, as one author titled his book, "The Devil's Backbone." That sinister designation fit the trail most compellingly, for it was a robber's lair where miscreants plied their evil trade with little let or hindrance. The trail was remote from civilization and open to the perfidy of bandits who would waylay well-heeled merchants returning northward from selling their merchandise in the port of New Orleans.

Even though some historians maintain that the trail had seen worse days than when Meriwether Lewis arrived on it in October 1809, still others, like historian Professor John Guice, have argued convincingly that the trail in 1809 was not yet a "yellow brick road" of safety and security for travelers. No wonder Mrs. Grinder waited until the coast was clear to give surcease to the dying pleas of Meriwether Lewis. Even in 1809, noninvolvement as a human reaction to danger prevailed. I don't think noninvolvement in the plight of others was what Christ meant by His adjuring mankind to turn the other cheek.

Other historians view Lewis's death as too cloaked in mystery to call it either a suicide or a homicide. Western historian Dee Brown has written that the manner of Lewis's death is mysterious. And John Bakeless, in his 1947 biography of Lewis and Clark, wrote, "The evidence for murder is not very strong, and the stories . . . strongly suggest suicide, but none of the evidence is really conclusive. It is impossible to make a positive statement either way."

Yet positive statements have been made, both ways. Stephen Ambrose wrote his historical novel *Undaunted Courage* promoting the suicide point of view. Contrariwise, Richard Dillon, in the first biography of Meriwether Lewis, calls his death a murder, not a suicide. And Ambrose's introduction to the paperback edition of Dillon's biography all but admits the credibility of Dillon's assertion. Ambrose there declares that "[t]he only reason I have not written his biography is that Richard Dillon did it first, and his is such a model biography there is no need for another."

And yet, just eight years later, Ambrose published his own suicide conclusion in his biography of Meriwether Lewis, leaving nary a glimmer of hope for any other position. And then, in a letter dated December 18, 1997, to President Clinton, Ambrose reiterated his view even more forcefully. He wrote: "What Starrs calls a controversy, isn't." Untroubled by his words in the introduction to Richard Dillon's biography, he there says he "never doubted that Lewis committed suicide . . . nor [does] any other serious scholar." Those words speak quite damningly of his uncompromising view of the contrary position of other serious historians, like Dee Brown and John Guice and even Richard Dillon. Will the real Stephen Ambrose please stand up?

Meriwether Lewis lies buried on land now owned and administered by the National Park Service along the Natchez Trace Parkway in Lewis County, Tennessee. The remains should be exhumed for scientifical analysis. The three accounts of Lewis's demise, although in many respects contradictory, nevertheless provide substantial avenues of scientific scrutiny for persons schooled in the forensic disciplines of forensic anthropology, forensic pathology, and forensic firearms identification. The results of these investigations may well contribute mightily to a resolution of the old and much-bruited question of whether Meriwether Lewis committed suicide or was the victim of a homicide.

I believe that to honor Lewis's bravery, genius, and contributions to this country, the historical record needs to be exposed to the cold, detached light of science. His bones might provide all the evidence needed to interpret his death definitively as homicide, suicide, or accident. One descendant of Lewis's sister, Jane, has even permitted her blood to be drawn and preserved for a later DNA comparison to ensure that the remains said to be those of Meriwether Lewis, when exhumed, are in fact his remains.

In consideration of the interaction of scientific team members representing the aforementioned disciplines, there is a reasonable probability that new factual insights into the manner of death of Meriwether Lewis can be discovered from his exhumed remains, as well as from associated artifacts in close proximity to the remains.

It is the peculiar forte of a forensic anthropologist to assay trauma to bone, also called "insult," whether occurring ante-, peri-, or postmortem. In light of the three accounts provided by Neelly, Wilson, and Russell, the anthropologist could focus on those bones that would be expected to yield evidence in support of or in opposition to these differing renditions. In view of the large caliber of the bullet or lead ball that characterized firearms in the early nineteenth century, and that Lewis was known to be carrying a brace of .69-caliber North and Cheney horseman's flintlock pistols, it would be anticipated that Lewis's skeletal remains would demonstrate the presence of insults to the bone, or lead wipe from a lead projectile, or even residues of black powder indicative of a contact or near-contact firing of a gun.

Not only the presence of such traces, but also the location and directionality of such impacts, would be exceedingly relevant to a determination of the manner of Lewis's death. Suppose, *ex hypothesi,* that an entrance wound were to be found in the rear of Lewis's skull. The length of the barrel of a .69-caliber North and Cheney would never—no matter how gymnastically adept a person might be—enable it to be fired in a suicidal manner in that fashion.

A pathologist could evaluate whether the bullet insults were of the defensive type, indicating their infliction by a third person. Recall the defensive wounds to bone suffered by Alfred Packer's victims. In addition, a pathologist could appraise the location and severity of bullet strikes to the bone to resolve the question of whether the varied accounts of Meriwether

Lewis's physical activities following his wounding were based on fact or fancy. The issue of the duration of a person's remaining conscious after the extravasation of significant amounts of blood is an important physiological curb on hyperbolic accounts of a person's movements after sustaining significant firearms injuries.

The forensic firearms examiner would seek evidence—minimally, from the presence or the absence of black powder residues—to support a conclusion concerning the distance from the gun to the target at the time of the firearm's discharge. This expert would also seek other black-powder-related materials, such as the spent projectiles, the patching material used to seat the lead ball in the barrel, and the imprint of the patch or patches on any lead balls recovered. The discovery of multiple lead balls and patches could be most instructive about the number of guns fired, the caliber of the guns discharged, and the ultimate question of the manner of Lewis's death.

Most assuredly, the possibility of a successful scientific investigation depends, in large part, on the condition of the remains at the site of interment. Prescinding from the data that could be derived from the artifacts, the answers to the essential questions concerning the death of Meriwether Lewis would be heavily reliant upon the state of preservation of the remains. All of my analyses to this point have given promise that the remains will be in sufficiently well-preserved condition to enable a scientific examination to bear fruit.

Little is known about the 1809 burial except for the vague and uninformative statement given by Major Neelly that he "had him as decently Buried as I could in that place." Yet in 1848, when the state of Tennessee erected the broken-shaft monument to Meriwether Lewis at the site of his death, the report of the committee appointed by the legislature indicated that "the grave was re-opened and the upper portion of the skeleton examined, and such evidences found leave no doubt of the place of interment."

Two important taphonomic facts are stated in this report: First, the coffin was still sufficiently intact to require reopening. Second, the remains were found to be skeletonized and sufficient for examination. However, the committee's report makes no mention of the method of their reburying the remains, whether in the original coffin or a new one. Nor did the committee, which included a medical doctor, provide even a glimmer of knowledge as to whether there were bullet wounds and where they might have been found. But they did, at least, describe erecting a monument over the new grave, which should be some considerable proof that Lewis's remains are under that monument.

This monument, with its nine-foot-square base, has likely shielded the grave from environmental ravages. In 1935, the National Park Service also placed four feet of backfill around its base, which has provided a further buffer zone for the remains from the elements while, ruefully, distracting from the picturesque symmetry and the elegant stature of the original monument.

It is known that from March 24 to April 2, 1998, the National Park Service's contract with a stonemason resulted in the temporary removal of the backfill for the purpose of determining the underground condition of the monument's base. I had previously informed the National Park Service's Tupelo headquarters of my having received the opinion of a Nashville professional engineer, Jack Wood, president of Barge, Waggoner, Sumner and Canon, Inc., that the base might be so unstable that the monument might topple over at any time. The Tupelo office was unwilling to credit my professional engineer's assessment and paid good taxpayers' money to a stonemason for the same purpose.

At the time of the stonemason's trenching, to a depth of four feet, I personally observed the backfill to be at all levels bone-dry. The removed soil, later replaced, bodes well for the good preservation of the remains beneath.

The pH of the soil (acidity versus alkalinity) and the soil's mineralogy (permeability for water runoff) were not testable by me from soil samples taken at the monument or anywhere else on National Park Service–owned lands per policy restrictions. I wondered whether they would monitor the soil I might unwittingly remove in the treads of my auto tires or the treads on the soles of my shoes.

More recently, in my investigations into the death of Billy the Kid I have been confronted with a similar hard-nosed attitude toward my soil sampling. The city of Silver City, New Mexico, has refused to allow soil sampling by me, no matter how paltry it might be, at the burial place of Billy the Kid's mother, Catherine McCarty, in the Memory Lane Cemetery. However, on this occasion, playing hardball has a different motivation. The refusal has been made in an attempt to frustrate my efforts to determine the scientific legitimacy of an exhumation of Catherine McCarty. I cannot say that the National Park Service was following a similarly small-minded path in the case of Meriwether Lewis.

As a result, I was relegated to going off park property to obtain a soil sample on land owned by ever-helpful local historian Marjorie Graves and her husband. The Graves (a felicitous surname if ever there was one) soil sample proved to be sufficient to relieve another anxiety over the preservation of the remains.

It is correct to say that my research had turned me into a Meriwether Lewis devotee. It was a warming experience to be investigating the death of a great man. Here was no Jesse James, no Albert DeSalvo, and certainly no Alfred Packer. Now, for the first time in my career as an exhumer, I had a notable and historic person of immense stature whose death, come what may from my work, demanded my unfettered attention. He had become my investigative passion, though I little realized the hurdles that would be planted ever higher in my path toward the true facts of his death.

Apart from a visit to his Tennessee grave and seeking to determine the soil constituents there to estimate the condition of his remains, the first order of my business was to obtain consents to an exhumation—for all sides and both far and wide. Locating living descendants of Meriwether Lewis was simply a matter of visiting the Hohenwald, Tennessee, public library, where the names of prominent descendants were recorded. Unlike the stone-faced attitude at the Fall River Historical Society, the librarian in Hohenwald was delighted to provide me with names and addresses. The name that she singled out as the one I should contact first was William M. Anderson, M.D., of Williamsburg, Virginia.

Upon returning to my office from my Tennessee trip, I immediately telephoned Dr. Anderson, who promptly and enthusiastically agreed to join forces with me to accomplish the objectives of an exhumation. From Dr. Anderson I was led to Jane Sale Henley, also a descendant of Jane Lewis Anderson, Meriwether Lewis's sister. Jane later became the president of the very influential Lewis & Clark Trail Heritage Foundation, which, courtesy of its former president, Arlen Large, passed a resolution in support of an exhumation of Lewis.

At this point, the stone of consents was proceeding smoothly downhill without gathering any moss. To date I have received signed formal consents to an exhumation from 170 persons claiming to be descendants of Meriwether Lewis, all in various collateral lines of descent, there being no known direct descendants of Meriwether Lewis. More official consents were received from the Lewis County Board of Commissioners due to the gentle lobbying efforts of a Hohenwald attorney, Tony Turnbow. The Lewis County Historical Society also passed a resolution signed by, of all names, its president, Philip Griner. And Terry Bunch, the Lewis County Executive, weighed in with a strong letter in support of me and the exhumation I proposed. All of these consents were unequivocal and unconditional, and as firm as firm could be.

Still not satisfied that my groundwork would convince the National Park Service to grant permission to exhume, I sought out the Lewis County District attorney general, Joseph Baugh, who was readily available and very game to assist me. Our discussions led us to agree that Joe Baugh would, in his official capacity representing the law-enforcement needs of Lewis County, direct the holding of a coroner's inquest into the death of Meriwether Lewis in 1809, since there was no record of a previous inquest and the death could reasonably be deemed to be suspicious.

That inquest, in Hohenwald, the Lewis County seat, was held for two days, on June 3 and 4, 1996, with Richard Tate, the local coroner, presiding over the receipt of testimony and the deliberations of seven jurors from the community, including one who was a medical doctor. After the National Park Service declined, at my invitation, to take part in the proceedings (although their Tupelo representative, Gary Mason, was in attendance as a spectator), I obtained the consent of fourteen experts in a variety of disciplines to appear and testify.

Among them was Dr. William Bass, a forensic anthropologist of considerable national note, as well as three historians, Arlen Large, Dr. John Guice, and Ruth Frick; one geophysicist, Dr. George Stephens; one psychologist, Dr. Thomas Streed; two pathologists, Drs. Jerry Francisco and Martin Fackler; a firearms expert, Lucien "Luke" Haag (who demonstrated for the jury the firing of a 1799 .69-caliber North and Cheney flintlock); an epidemiologist, Dr. Reimert Ravenholt; and two document examiners, Dr. Duayne Dillon and Gerald Richards, who spoke to the issue of what Meriwether Lewis's handwriting could tell us of his state of mind when the documents were drafted—if they were drafted or signed by him.

The testimony of all of the experts was presented through the questioning of Joe Baugh and through the queries of the experts from the members of the coroner's jury. It was a most impressive array of talent and a most telling presentation by all and sundry. The result, as announced by the coroner, Richard Tate, was a unanimous verdict signifying that "there is very little tangible evidence for this jury to base a credible ruling as to the matter of suicide." As a result, it was stated, "because of the importance of the person in question to the history of Lewis County, we feel exhumation is necessary for closure."

With the coroner's jury's verdict backing him, Joe Baugh, again in the role of the local prosecuting official, petitioned the Lewis County Circuit Court for an exhumation of Meriwether Lewis's remains. Everything augured well for an exhumation until the National Park Service intervened and removed the petition from the local state court to the federal district court in Nashville. At this juncture, the proceedings took a different and dispiriting turn.

The National Park Service, represented by the Nashville United States Attorney's office, now took the upper tactical hand. Joe Baugh was ousted from the more familiar precincts of the Tennessee state courthouse, where he was a frequent entrant, and thrust into a federal courthouse whose personnel, accoutrements, and procedures were not his accustomed fare, all through the legal pleadings of the U.S. Attorney. Little did Joe realize how much more distressful his comfort level would become when the day of his federal courtroom appearance before Federal District Judge Thomas Aquinas Higgins arrived.

As I assumed a spectator's seat in the courtroom, in the failing hope of being called by Joe to testify in support of his petition, the proceedings commenced with Judge Higgins presiding. No, not presiding but commanding and dictating the tone and tenor of the hearing.

Even before Joe could be given the time to explain the nature of his petition and his reasons for filing it, Judge Higgins stopped him in his tracks with a stinging rebuke on the spurious nature of the entire matter, from the judge's perspective, that is. The judge took Joe to task for exercising poor judgment in bringing this petition when, in the judge's view, Joe had more compelling current law-enforcement business to occupy his time. Joe attempted to gain and hold a foothold of rational argument, but Judge Higgins would have none of that, besieging him with the alleged inanity, as the judge perceived it, of the entire matter. It was as if I were seated in the classroom of the overbearing and arrogant Ivy League law Professor Charles W. Kingsfield, Jr., as he was depicted in the television series *The Paper Chase*. However, on this occasion it was fact, not fiction, that was meanly on display.

The judge's scornful comments settled two issues for me: that his mind was not open to any arguments from Joe for an exhumation, no matter how well conceived and presented, and that the petition was, in the judge's mind, a lost cause even before the assistant U.S. Attorney was heard in opposition to it.

And so, predictably, Judge Higgins's frame of mind at the oral argument was transferred into his 1998 published opinion rejecting the merit of the request for an exhumation. After such a demeaning experience in federal court, it was no wonder that Joe decided against appealing Judge Higgins's decision. Once burned; forever fear the flame.

Having exhausted my efforts in the courts, state and federal, I was flung headlong into a new and unfamiliar arena, the maelstrom of government agencies and political rough-and-tumble. In June 1997, sensing in advance what Judge Higgins's written opinion would say, I joined with Dr. William Bass, the nation's premier forensic anthropologist, in filing an application under the federal statute known as the Archaeological Resources Protection Act of 1979 (ARPA) to obtain a permit granting us authority to exhume Meriwether Lewis. We were now within the discretionary clutches of the National Park Service (NPS) of the Department of the Interior, which held ownership and stewardship of Lewis's burial place. The act was controlling, since the remains were on federal land and they had been in place for more than a hundred years, thereby becoming an archaeological resource to be protected from any intrusion not allowed under the terms of the statute.

At first things went smoothly, with a memorandum of August 1, 1997, from Bennie Keel, the regional archaeologist for the National Park Service charged with reviewing the application, stating that the "technical merit" and the "issue of public interest" and "the education, experience, and training requirements" of both Dr. Bass and me were all satisfactory. There was nothing in this memorandum that mentioned or implied anything but an honest and fair appraisal of the project's merits.

Later it was learned, to the contrary, that the higher-ups in the NPS had already decided determinedly to oppose the application. On October 2, 1996, more than six months before the application was even filed, Gary Mason, representing Natchez Trace superintendent Daniel Brown, had forwarded a memorandum concerning a forthcoming October 4 planned "meeting . . . on M. Lewis."

That memo, in no uncertain terms, stated that the die was cast and the decision was made to deny my application to exhume, if and when it was filed. The words of this one-page memo left no room for doubt: "Director Kennedy, Stevenson and Pitcaithley remain steadfastly opposed to permitting the exhumation of ML," it said. It also, unshamefacedly, admitted to a cabal already in action "on setting aside the exhumation proposal." The plotters and schemers were stated to be "Pitcaithley and Kate Stevenson" in concert with Congressman Ed Bryant, the then representative for Lewis County.

"Kate Stevenson" was then the associate director of NPS in charge of Cultural Resource Stewardship and Partnerships, in essence the final voice on the application before it reached the NPS director. Even though in a meeting with her in the congressional offices of Representative Bob Clements of Nashville, a man whose statesmanship and congeniality were to me beyond reproach, she had represented her impartiality on the merits of my plans for an exhumation, still this Gary Mason memo now confirms that that claim to fairness was a misleading façade. Further assurance of that came later in the processing of the application, through the NPS chain of command.

The application's having been denied by Superintendent Daniel Brown at the Natchez Trace Headquarters and denied again, after a very fair and thorough hearing at the Atlanta Regional Office of the NPS, by Regional Director Jerry Belson, I had no alternative but to appeal to NPS Director Robert Stanton. It was then that Associate Director Katherine Stephenson's insider role became a matter of public record.

At the level in my quest that I had now reached, I realized that without legal assistance I would not be up to the tasks ahead. As a result, I asked Tim Means, a member of the Washington, D.C., law firm of Crowell & Moring and a well-regarded former student of mine at The George Washington University Law School, if he and his firm would honor me with their pro bono legal services. True to the Tim Means I remembered well, he and a number of senior members of his firm agreed to pilot the application through the appeal-review process with Director Stanton.

As my lawyers and I waited for answers from Director Stanton concerning the procedural details of his review, I received a phone call, later reproduced in a letter of April 28, 1998, to Director Stanton, from Steven P. Quarles, who was spiriting my efforts on behalf of Crowell & Moring. The call and the letter following it were enough to make a saint a sinner or a naive professor a confirmed cynic, for it proved, without a shadow of a doubt, that the application to exhume had been prejudged in the backrooms of the NPS not on its merits but for some other reason that never really has seen the light of day.

Steve Quarles's April 28, 1998, letter to Director Stanton gives the factual details. At an NPS award ceremony one week earlier, he had been engaged in a conversation with a Ms. Beller, apparently a client of Crowell & Moring, about his current legal activities. When he mentioned his providing legal representation in behalf of the appeal of my application for an exhumation of Meriwether Lewis, Ms. Stevenson, cultural affairs associate director of the NPS, "abruptly entered the conversation." Ms. Stevenson openly stated "her hostility to the proposed project and her absolute opposition to" my application. She topped her views off by stating that the "application would be granted only over my [Ms. Stevenson's] dead body."

In light of Ms. Stevenson's statements and her having, on the following day, telephoned an official in a firm represented by Crowell & Moring in its dealings with the NPS, complaining "about [his] representation of

Professor Starrs['s]" application, Steve Quarles quite properly asked Director Stanton to recuse Ms. Stevenson from the director's previously announced role as a reviewer of my appeal.

When Steve Quarles telephoned me with this wretched information, I was at a temporary loss for words until he referred to Ms. Stevenson's saying "the application would be granted only over my dead body." My reply was simply to ask if he had told her that's how I work best, over dead bodies—not hers, of course. The outcome of the recusal request gave no promise of the application's being fairly heard on appeal. By a letter dated July 7, 1998, Director Stanton denied the request, saying Ms. Stevenson's views would not be determinative but that he had the final decision—which, of course, was nothing more than saying he would base his opinion on a marked card in a stacked deck.

The inevitable outcome came as a matter of course in a letter from Director Stanton with an accompanying memorandum from his staff stating, in part, that Dr. Bass and I were unqualified to lead such a project and that The George Washington University was not suitable as a repository of any artifacts discovered in the exhumation, rejecting GWU vice president Dr. Carole Sigelman's letter agreeing to be a repository. Other, similarly specious arguments were for the first time presented, leaving us no opportunity to respond or to be granted reconsideration.

So ended my first and only venture into governmental dealing and double-dealing on behalf of a most worthwhile exhumation to provide a voice for the remains of a man justly revered as our nation's premier explorer. The voices of 170 descendants of Meriwether Lewis have not been heard, nor have those of three governors, Virginia's George Allen, Tennessee's Don Sundquist, and Missouri's Mel Carnahan, all of whom put their voices in writing in support of the exhumation application. And in spite of a multidisciplinary scientific team representing ten different colleges or universities, as well as others in the private sphere, the application fell on deaf ears. What could account for this out-and-out rejection even before the application to exhume was formally filed?

I put aside as inconceivable that the NPS could have reacted in a niggardly or vengeful way to the discoveries by me in the course of my investigations that could only embarrass the NPS in its stewardship of the

Lewis grave site. Three such finds were particularly grating and good cause for public concern over the failure of the NPS to live up to the terms of its legislative stewardship responsibilities, as was previously the case with the decaying and decrepit condition of Grant's Tomb in New York City, also under NPS stewardship.

First I was caught up in my preliminary investigations with securing pictures of the Lewis monument as it was over the years. I was preparing a monograph on the monument that required a series of photos of the monument then and now. But I discerned that something was amiss, particularly when I compared the monument of today to that when former Tennessee governor Austin Peay was pictured there in the 1920s, as well as when Meriwether Lewis Anderson, M.D., great-grandnephew of Meriwether Lewis, was photographed at the Lewis monument in 1933.

The short of it was that the monument today was four feet shorter than it was when erected in 1848 by Columbia, Tennessee, stonemason Lemuel Kirby. The base of the monument was, according to the Monument Commission's report in 1848, eight feet high, but today it was only forty-four inches high. Where had the other four feet gone? My digging in the historical records proved that sometime in the middle 1930s, under the stewardship of the federal government, four feet of backfill had been mounded around the base of the monument, obscuring four feet of its solitary original presence. When I informed Superintendent Daniel Brown of this discovery, he was at first in disbelief over such a blatant derogation of this national monument, but, as he said, it was not done on his watch.

A second embarrassing moment occurred when the ground-penetrating radar (GPR) runs conducted by my scientific team, with the consent of Superintendent Brown, at the site in 1992 resulted in another distressing find. The grid laid out by George Stephens, Paul Baldauf, doctoral candidate in the Geology Department at The George Washington University, and me for the GPR was broad enough to encompass a wide swath of smallish ground-level burial markers on the periphery of the Lewis monument supposedly bearing the surnames of the pioneers buried there.

However, according to the results reported by Dr. Jim Mellett, a geophysicist affiliated with New York University, a sizable number of the markers were over plots where there was no GPR evidence of an anomaly provable of a burial there. Worse yet, the printouts of the runs revealed places where there were underground anomalies without named grave markers above them. This was a disturbing find that led me to inquire when the markers had been placed at the site as replacements for the decaying monuments that clearly appear in old pictures taken there.

It appears that the plain, ground-level markers were set in place in 1928 as part of Superintendent DeLong Rice's refurbishing of the monument and the area surrounding it. But in 1935, when the backfill arrived at the scene, at least some of the markers were either buried under the soil or replaced after the soil had been mounded around the monument. If after the backfill was in place, it is altogether likely that the markers were haphazardly relocated, possibly at sites where they did not belong, as we found from our GPR runs.

It was the 1928 refurbishing and removal of the decaying monuments at the site that created the groundwork for the third embarrassing moment for the NPS. The monuments, according to the recollections of persons living at the time, were unceremoniously bulldozed off the site. As if that were not indignity enough, they were stored leaning up against a maintenance shed on the property until I took pictures of their dismal location in 1997 and gave public notice of this disgrace, with pictures of them, in my monograph titled *Meriwether Lewis: His Death and His Monument.*

Sad reflections on the NPS's stewardship over the Lewis monument that these three findings entailed, still I sensed that my disclosures of them were insufficient to cause the rock-hard and intransient refusal to grant me a permit for an exhumation. Probing deeper in the political underworld, I came to a realization of the likely truth of the matter.

My search for support for my project inevitably led me to Capitol Hill in Washington, D.C., and the offices of Lewis County congressman Bill Bryant and Nashville's Congressman Bob Clements. I never managed to rate a personal audience with Congressman Bryant, but his staff member, Mark Johnson, made it quite clear to me, with his churlish manner bordering on the arrogant, that I was not welcome at the congressman's table. I then sought the succor of Congressman Clements and found him to be gracious and receptive to my overtures for support, a world apart from the cold freeze with Bryant's staff. Bob Clements was inspired enough by my pleas to host a meeting in his offices with Katherine Stevenson of the NPS as well as the staff of a number of politicos from Tennessee. At that meeting, and upon my pressing for the basis for the NPS antagonism to my proposal, Ms. Stevenson said that I was not the reason for the NPS's refusals, but she did not go further than that.

It was from a reading of the Gary Mason memo of October 2, 1996, that it all came clear. In that memo, Mason makes two points that are relevant at this juncture. In paragraph number 1, he declares that Congressman Bryant had been meeting with NPS representatives Pitcaithley and Kate Stevenson to arrange a scheme to sandbag my project. At the same time, Mason speaks of "setting aside the exhumation proposal while exploring other bases for bringing attention to Lewis County and its tourism potential."

Now the cards were on the table, and for me and my project they were all jokers. I was trapped in the middle of a governmental byplay, with the NPS being importuned by Congressman Bryant to support his pet pork project on the Natchez Trace and the NPS putting my project up as a bargaining chip. In the world behind the scenes of governmental dealings, my project just didn't have the staying power. So just as it was the Washington governmental bureaucracy that set Meriwether Lewis on his last and fatal journey, it was the same bureaucracy that had frustrated my efforts to set the record straight on the death of Meriwether Lewis. There is a certain perverse irony in that.

But I have not turned tail, nor am I resigned to the NPS rejection of my application to exhume as being the end of the affair. In the words of Edna St. Vincent Millay, "I do not approve. And I am not resigned." But what's to be done? Is there any hope that an exhumation will ensue? I believe that I can state, without the optimism of naïveté, that the hope of such an outcome is much more than the "hope that springs eternal." It is the hope of a realist, not a visionary.

There are other avenues of recourse beyond those provided by the Federal Archaeological Resource Protection Act. Speaking rhetorically, if the rights of burial of Native Americans can be protected under federal law, even when the burial is many centuries old or more, as in the case of the nine-thousand-year-old skeletal remains of the Kennewick Man from Washington State, are not the remains of Meriwether Lewis entitled to the same rights in law? This question poses a how-to-do-it for me and for Lewis's determined relatives, led by the energy and enthusiasm of Howell Bowen and Jane Henley, descendants of Meriwether Lewis's sister.

One approach would be that of new federal legislation giving the descendants of Meriwether Lewis the final incontestable say over the decision to exhume Lewis's remains. Such legislation would guarantee an exhumation, the descendants being solidly and just shy of 100 percent in support.

Another plan would involve reliance upon litigation founded on common-law and constitutional-law principles relating to the basic right of the descendants to have control over the remains of their ancestor, Meriwether Lewis. Such litigation could justifiably lay claim to the universal right, enshrined in the United Nations Declaration of Human Rights, to a burial in accordance with the religious dictates of the deceased's family and descendants. Just as there was no documentation of any inquest having been conducted in 1809 into the death of Lewis, so too there is no record, oral or otherwise, that Lewis received the solemn ceremony of a Christian burial. I do not believe that the lapse of time has dimmed or forsworn that rite of passage to him, which can only occur after an exhumation has confirmed his identity at his present burial place.

Another avenue of litigation would be that provided by the Federal Advisory Committee Act of Title 5 of the United States Code. It is more than merely arguable, under this act, that the National Park Service's having taken into consideration in its decision-making the recommendations against exhumation of Stephen Ambrose and other private persons without permitting an open hearing on the issue to all comers would void the action it took in denying my exhumation. In that event, the process of review would be reinstituted with a new Park Service director and Ms. Stevenson no longer an assistant director at the Park Service. The presence of new personnel in controlling positions at the highest levels of the Park Service gives me the real hope of carrying the day for an exhumation that previously had only been devoutly wished for.

In short, the effort to exhume Meriwether Lewis and to resolve the manner of his death is ongoing, is a work-in-progress for me, my team members in science and history, as well as his descendants. With such enthusiastic and determined support, I am convinced that the exhumation of

Meriwether Lewis will become a reality rather than a dream. After all is said and written, all that remains to answer this historical enigma are Lewis's remains.

The bicentennial of the Lewis and Clark expedition is presently under way. From 2003 to 2006, attention will be directed to every aspect of the lives of Meriwether Lewis and William Clark. Lewis's mysterious death will be on the agenda for open discussion in the many venues across the land where his leadership of the Corps of Discovery will be headlined as the topic *du jour.* An article by two historians, J. Frederick Fausz and Michael A. Gavin, in the 2004 winter edition of *Gateway Heritage,* the quarterly magazine of the Missouri Historical Society, puts the case for an exhumation quite simply and forcefully: "Rather than being disrespectful, [an exhumation] would be the most appropriate and meaningful bicentennial tribute to the famous explorer, reflecting his own commitment to employing advanced scientific expertise in making new discoveries."

A national hero of the stature of Meriwether Lewis deserves no less in death than the plaudits he received in life. It is my steadfast objective to see that no effort is too great nor any is deemed wasted in seeking to clarify the manner of his death.

IN ALL OF MY EFFORTS IN SEEKING TO PROVIDE A VOICE FOR THE dead, I have been guided by the words of Voltaire, who said: "We owe respect to the living; to the dead we owe only truth."

It is well to take heed that death does not wipe life's slate clean. As poet laureate John Masefield tersely put it:

For death takes toll
Of beauty, courage, youth,
Of all but truth.

My voice, therefore, speaks for the dead on behalf of the living, the living whose suffering is real, tangible, and waits for science to relieve.

It is my hope that the chapters of this book have convinced the reader, who will carry the message forth, that exhumations are not a walk in the park, a jog around the corner, or even a tumble in the hay. Exhumations take the coordinated efforts of many disciplines, both scientific and otherwise. They are also, when conducted in the private sphere, time-consuming, costly, and fraught with unexpected weirs at every turn.

But there are rewards beyond measure. When Diane Dodd says to me that "God and Mary will forgive you" for caring for her sister, Mary Sullivan, rather than my Sunday Mass obligation, I am humbled and chastened. When I see students working ceaselessly in the blazing sun at the screens seeking fragments of bones and artifacts for the benefit of the historical record, I am cheered in the knowledge that the life I have given over to teaching has not been in vain.

When I work in the company of competent and skilled scientists who freely give their labors in a collaborative enterprise so that the secrets of the dead can be unlocked, I realize that there is goodness in mankind that knows no bounds. And when I receive my annual box of Florida oranges from the family of Carl Weiss, M.D., I know, as in the case of the rose that annually makes its appearance at the grave of Edgar Allan Poe, that people are thankful and do not forget.

Can anyone ask for more, unless it be, in my case, another deserving exhumation?

SELECTED BIBLIOGRAPHY
BY CHAPTER

INTRODUCTION

Guttridge, Leonard F. "Identification and Autopsy of John Wilkes Booth," *Navy Medicine*, January–February 1993.

Kline v. Greenmount Cemetery, 677 A. 2d 623 (Md. App. 1996).

True Crime. *Assassination*. Richmond, Va.: Time-Life Books, 1994.

CHAPTER ONE

"A Colorado Tragedy," *Harper's Weekly*, Oct. 17, 1874.

Fenwick, Robert Wesley. "Alfred Packer: The True Story of the Man-Eater," *Denver Post*, 1963.

Gantt, Paul H. *The Case of Alfred Packer: The Man Eater*. Denver: Denver University Press, 1952.

Hodges, J. G. "The Legal Experience of Mr. Alfred Packer," *Dicta*, vol. 19 (Denver, June 1942).

Jocknick, Sidney. *Early Days on the Western Slope of Colorado*. Glorieta, N.Mex.: Rio Grande Press, 1913.

Kushner, Ervan F. *Alfred G. Packer: Cannibal! Victim?* Frederick, Colo.: Platte 'N Press, 1980.

Mazzula, Fred and Jo. *Al Packer: A Colorado Cannibal.* Denver, Colo.: Fred and Jo Mazzula, 1968.

McClain, Joni L., Fred Jordan, and Ray Blakeney. "Human Cannibalism: A Case Report," *Am. J. For. Med. & Path.* 7, no. 2 (1986): 172–173.

Packer v. People, 8 Pac. 564 (Colo. 1885).

In re Packer, 33 Pac. 578 (Colo. 1893).

Packer v. People, 57 Pac. 1087 (Colo. 1899).

"Packer, the Ghoul," *Silverworld,* Colo., March 3, 1883, p. 3.

"The Packer Case," *Silverworld,* Colo., March 24, 1883, p. 3.

"Packer at Lake City," *Silverworld,* Colo., March 31, 1883, p. 3.

Rogers, Will. "Given Forty Years in Jail, He Was Let Out When Republicans Needed Some More Opponents Devoured in Close Election," *San Francisco Chronicle,* November 2, 1930.

Wise, Joe. *Cannibal Plateau.* Sante Fe, N.Mex: Sunstone Press, 1997.

CHAPTER TWO

Deutsch, Hermann B. *The Huey Long Murder Case.* Garden City, N.Y.: Doubleday & Company, Inc., 1963.

Goodman, John. *Kingfish: A Story of Huey P. Long.* Turner Pictures Video, 1995.

Hair, William Ivy. *The Kingfish and His Realm: The Life and Times of Huey P. Long.* Baton Rouge: Louisiana State University Press, 1991.

Heleniak, Roman. *Soldiers of the Law: Louisiana State Police.* Topeka, Kans.: Josten's Publications, 1980.

Jolly, Ellen Roy, and James Calhoun. *The Louisiana Capitol.* Gretna, La.: Pelican Publishing Company, 1980.

Reed, Ed. *Requiem for a Kingfish.* Baton Rouge, La.: Award Publications/Ed Reed Organization, 1986.

Starrs, James E. "Scientific Insights into a Louisiana Tragedy," *Sci. Sleuth. Rev.* 15, no. 3 (Summer 1991): 1–12.

True Crime. *Assassination.* Alexandria, Va.: Time-Life Books, 1994.

Williams, T. Harry. *Huey Long.* New York: Alfred A. Knopf, 1969.

Zinman, David. *The Day Huey Long Was Shot.* Jackson: University Press of Mississippi, 1993.

CHAPTER THREE

Albarelli, H. P., Jr. "The Mysterious Death of CIA Scientist Frank Olson," www.crimemagazine.com/olson.htm; and www.crimemagazine.com/olson2.htm.

Desmon, Stephanie. "In Reburial, Olsons Hope to Lay Saga of Father to Rest," *Baltimore Sun,* August 9, 2002.

Dowling Kevin, and Phillip Knightley. "The Olsen File: A Secret That Could Destroy the CIA," *Night and Day,* August 23, 1998.

Fischer, Mary A. "The Man Who Knew Too Much," *Gentleman's Quarterly,* January 2000.

Gup, Ted. "The Coldest Warrior," *Washington Post,* December 16, 2001.

Ignatieff, Michael. "What Did the CIA Do to Eric Olson's Father?" *New York Times,* April 1, 2001.

Marks, John. *Chapter 5: The Search for the Manchurian Candidate: The CIA and Mind Control.* New York: Times Books, 1979.

O'Toole, Thomas. "Suicide Revealed: CIA Infiltrated 17 Area Groups, Gave Out LSD," *Washington Post,* June 11, 1975.

Szulc, Tad. "The CIA's Electric Kool-Aid Acid Test," *Psychology Today,* November 1977.

Vankin, Jonathan, and John Whalen. *Chapter 80: Acid Drop: The Greatest Conspiracies of All Time.* Citadel Press, 2004.

www.frankolsonproject.org

CHAPTER FOUR

Breihan, Carl W. *The Complete and Authentic Life of Jesse James.* New York: Frederic Fell, 1953.

Crey, Homer. *Jesse James Was My Neighbor.* New York: Duell, Sloan and Pearce, 1949.

Hansen, Ron. *The Assassination of Jesse James.* New York: HarperCollins, 1983.

Love, Robertus. *The Rise and Fall of Jesse James.* New York: Blue Ribbon Books, 1939 (reprint of 1926 G. P. Putnam's Sons edition).

Settle, William A., Jr. *Jesse James Was His Name.* Lincoln: University of Nebraska Press, 1977.

Stiles, T. J. *Jesse James: Last Rebel of the Civil War.* New York: Alfred A. Knopf, 2002.

Triplett, Frank. *The Life, Times and Treacherous Death of Jesse James.* New York: Swallow Press, Inc., 1970 (reprint of 1882 edition).

Yeatman, Ted P. *Frank and Jesse James.* Nashville, Tenn.: Cumberland House, 2000.

CHAPTER FIVE

Bailey, F. Lee, and Harvey Aronson. *The Defense Never Rests*. New York: Stein and Day, 1971.

Frank, Gerald. *The Boston Strangler*. New York: Penguin Books, 1967.

Hearst, Patricia Campbell. *Every Sweet Thing*. New York: Pinnacle Books, 1982.

Kelly, Susan. *The Boston Stranglers*. New York: Kensington Publishing, 1995, 2002.

Rae, George William. *Confessions of the Boston Strangler*. New York: Pyramid Books, 1967.

Sherman, Casey. *A Rose for Mary: The Hunt for the Real Boston Strangler*. Boston: Northeastern University Press, 2003.

CHAPTER SIX

GOUVERNEUR MORRIS

Brookhiser, Richard. *Gentleman Revolutionary: Gouverneur Morris—The Rake Who Wrote the Constitution*. New York: Free Press, 2003.

LIZZIE BORDEN

Flynn, Robert A. *The Borden Murders: An Annotated Bibliography*. Portland, Me.: King Philip, 1992.

Hoffman, Paul D. *Yesterday in Old Fall River: A Lizzie Borden Companion*. Durham, N.C.: Carolina Academic Press, 2000.

Kasserman, David R. *Fall River Outrage: Life, Murder, and Justice in Early Industrial New England*. Philadelphia: University of Pennsylvania Press, 1986.

Kent, David. *The Lizzie Borden Sourcebook*. Boston: Branden, 1992.

Lincoln, Victoria. *A Private Disgrace: Lizzie Borden by Daylight*. New York: G. P. Putnam's Sons, 1967.

Pearson, Edmund, ed. *Trial of Lizzie Borden*. Birmingham, Ala.: The Notable Trials Library, 1989.

Phillips, Arthur S. *The Borden Murder Mystery: In Defence of Lizzie Borden*. Portland, Me.: King Philip, 1986.

Porter, Edwin H. *The Fall River Tragedy: A History of the Borden Murders*. Portland, Me.: King Philip, 1985.

Spiering, Frank. *Lizzie*. New York: Random House, 1984.

Sullivan, Robert. *Goodbye Lizzie Borden*. Brattleboro, Vt.: The Stephen Greene Press, 1974.

Ambrose, Stephen. *Undaunted Courage: Meriwether Lewis, Thomas Jefferson, and the Opening of the American West.* New York: Simon & Schuster, 1996.

Bakeless, John. *Lewis & Clark: Partners in Discovery.* New York: William Morrow & Company, 1947.

Daniels, Jonathan. *The Devil's Backbone: The Story of the Natchez Trace.* Gretna, La.: Pelican, 1989.

Dillon, Richard. *Meriwether Lewis: A Biography.* New York: Coward-McCann, 1965.

Dillon, Richard. *Meriwether Lewis: A Biography.* Santa Cruz, Calif.: Western Tanager Press, 1988.

Fausz, J. Frederick, and Michael A. Gavin. "The Death of Meriwether Lewis: An Unsolved Mystery," in *Lewis & Clark: New Perspectives,* special issue of *Gateway Heritage.* (Missouri Historical Society) vol. 24, no. 2–3, (2003–2004).

Fisher, Vardis. *Suicide or Murder? The Strange Death of Governor Meriwether Lewis.* Denver: Alan Swallow, 1962.

In Re Exhumation of Lewis, 999 F. Supp. 1066, M.D. Tenn., 1998.

Jackson, Donald. *Letters of the Lewis and Clark Expedition with Related Documents, 1783–1854,* 2nd ed. Urbana: University of Illinois Press, 1978.

Starrs, James E. *Meriwether Lewis: His Death and His Monument.* Washington, D.C.: The George Washington University, 1997.

INDEX